D1823320

In-situ concrete industrial hardstandings

Specification, design, construction and behaviour

John Knapton

Thomas Telford

Published by Thomas Telford Publishing, Thomas Telford Limited,
1 Heron Quay, London E14 4JD.

URL: http://www.t-telford.co.uk

Distributors for Thomas Telford books are
USA: ASCE Press, 1801 Alexander Bell Drive, Reston, VA 20191-4400
Japan: Maruzen Co. Ltd, Book Department, 3–10 Nihonbashi 2-chome, Chuo-ku, Tokyo
103
Australia: DA Books and Journals, 648 Whitehorse Road, Mitcham 3132, Victoria

First published 1999

A catalogue record for this book is available from the British Library

ISBN: 0 7277 2827 X

Typeset by MHL Typesetting Ltd, Coventry.
Printed and bound in Great Britain by Bookcraft (Bath) Ltd.

Preface

Although in-situ concrete is used commonly as the construction material for industrial hardstandings, to date there has been no 'one-stop' source of detailed guidance for designers, specifiers and constructors. To date, hardstandings have been engineered by merging together the methods and materials used in highways and industrial ground supported floors. As a result, industrial hardstandings have frequently been constructed using ineffective materials, and the detailing of joints and the introduction of reinforcement have been inappropriate. A small number of specialist designers has emerged whose expertise is based upon their experience rather than upon sound engineering principles. Other designers often prefer to use alternative less suitable materials for hardstandings.

The need for a rigorous methodology has heightened following the introduction of heavy container handling equipment which may apply wheel loads an order of magnitude greater than those applied by highway vehicles. Straddle carriers and front lift trucks that handle containers need to be fully quantified in terms of wheel loads, dynamic factors and spectrum of movements in order that a correctly engineered pavement design can result. Also, such equipment can lead to significant levels of stress in deep underlying subgrade materials, even when the attenuating effect of the concrete is taken into account. For this reason, a rigorous analysis of the geotechnical factors needs to be undertaken.

Recent changes in the consumer products distribution industry have led to a need for larger hardstandings such that the cost of the hardstanding may be a significant proportion of the cost of a development. In such cases, therefore, refining the design of the hardstanding may influence the cost of the project significantly. The introduction of new materials such as polypropylene and steel fibres has given the designer further opportunities to match the hardstanding to the needs of the user whilst at the same time creating more opportunities for errors.

Details of Author
John Knapton is Professor of Structural Engineering at the University of Newcastle upon Tyne where he pursues his research interests into

industrial pavements, heavy duty paving, pavements surfaced with pavers, highway pavements, port pavements and aircraft pavements. Before taking the Newcastle Structures chair, he ran his own consulting practice and was involved in the design and construction of many industrial hardstandings. His early career included spells as the British Constructional Steelwork Association Research Associate at Newcastle, Research Engineer in Cement & Concrete Association's Construction Research Department at Wrexham Springs and Lecturer in Structural Engineering at Newcastle. He has represented the Institution of Civil Engineers on BSI committees and has published over 100 papers since 1974. His work in developing the Ghanaian rural village of Ekumfi-Atakwa has led to his enstoolment as Nana Odapagyan Ekumfi I (Chief Eagle of Ekumfi I) by the Ghanaian government's House of Chiefs.

Contents

Introduction

In-situ concrete hardstandings are commonly constructed in industrial societies and are frequently the single most expensive element of a civil engineering project. Unlike other concrete elements, there is little guidance available to the designer — design, specification and construction methods developed for highway pavements or industrial floors are generally adapted. This usually leads to inefficiencies in the use of materials and has in some cases led to failure.

Because failure of an industrial hardstanding is less dramatic than failure of other concrete elements, little attention has been paid to the development of appropiate technology. However, there are unique issues to be addressed. For example, compared with a highway, the loading can be an order of magnitude greater. Most industrial hardstandings have significant length in two orthogonal directions, whereas a highway pavement often has a width of say 10 m. This leads to a need to introduce an arrangement of joints which allows the hardstanding to accept temperature- and moisture-related movements in two (and a half) dimensions without generating untoward stresses. It also requires a different approach to planning the construction work.

If a concrete hardstanding were unrestrained, it would be able to contract or expand in an unimpeded fashion so no stresses would be generated. The slab/sub-base interface generates frictional resistance to expansion or contraction and so induces stresses. Moisture or temperature variation through the depth of a slab results in the slab's attempting to curl — either upwards at the edges when the underside is wetter or warmer than the upper surface or downwards at the edges when the reverse is true. Self-weight and imposed loading prevent such curling and so introduce a further pattern of stresses. All of the stresses created by restraint to volume change can be managed either by introducing joints or by including reinforcement in the concrete (or by a combination of the two). For example, many unreinforced concrete industrial hardstandings include an orthogonal arrangement of joints on a 5 m grid. Each 25 m^2 bay is an independent structure requiring its own design and specification. Some of the independence is frequently removed by the introduction of dowel bars or ties to limit the relative movement between neighbouring bays or to ensure that neighbours share the burden of withstanding imposed loading.

As an alternative to joints spaced at 5 m, steel reinforcement can be introduced into the concrete. Such reinforcement does not prevent cracking, rather it transforms an occasional major crack into a series of regularly spaced unharmful narrow cracks — the more the reinforcement the greater the number of cracks. Reinforcement in in-situ hardstandings should be thought of as crack management reinforcement rather than stress management reinforcement as in the case of traditional reinforced concrete. Steel reinforcement has traditionally taken the form of sheets comprising bars welded together in two orthogonal directions, known as mesh. More recently, steel fibres have been introduced which can be mixed into the concrete, either at the construction site or at a mixing plant. Steel fibres not only govern joint spacings in the same way as mesh but additionally increase the strength of the concrete so permitting the use of thinner slabs.

Other materials are sometimes included in the concrete. For example, polypropylene fibres can reduce many of the difficulties faced by constructors before the concrete sets and they can toughen the concrete. Air is sometimes deliberately introduced (or 'entrained') into the concrete to safeguard the concrete against damage by frost. Other additives can change the workability of the concrete.

In traditional long strip in-situ concrete hardstanding construction, slab thickness is governed by loads, concrete strength and ground conditions whilst the joint arrangement is governed by the type and quantity of mesh reinforcement. With steel fibre reinforced concrete, both slab thickness and joint configuration depend upon concrete properties.

The first two chapters of this book present an overview of hardstandinging materials and construction methods. In particular, polypropylene and steel fibres are described since it is common for them to be included in industrial hardstandings. Working methods, specification data and design strengths are provided for commercially available fibres. The spacings and details of joints are included here since they are considered to fall into the category of construction rather than design

Chapter 3 reviews some of the categories of loading commonly occurring on industrial hardstandings and deals in particular with the effects of container handling and storage systems on industrial hardstandings. A design chart is included in Chapter 4 which allows in-situ concrete hardstandings to be proportioned. The chart introduces the concept of 'single equivalent wheel load', whereby combinations of point loads of differing values and contact areas can be converted into a single load value and then used to specify slab thickness. Illustrative examples of design for point loading and patch loading are explained.

Chapter 5 presents a case study and considers the difficulties that can arise with hardstandings. It is based upon the Author's own investigations and construction projects. It includes a case study illustrating construction techniques. Chapter 6 comprises a full set of contract documents which the reader may wish to use as a basis for his own project. For convenience, they can be obtained electronically and Chapter 6 provides details of how to aquire them.

1. Materials

1.1 Concrete

1.1.1 Introduction

Concrete is a man made composite material comprising natural aggregate, water and cement. Other materials may be added to enhance the performance of the concrete. For most applications, concrete is defined or specified by its 28 day characteristic crushing strength, cement content and free water/cement ratio. When determining the load bearing capacity of an industrial hardstanding, the flexural strength of the concrete is needed. It is therefore necessary to relate flexural strength to the characteristic crushing (or cube) strength of concrete. The main structural element of an industrial hardstanding is the concrete slab which is constructed over the sub-base and subgrade material, and careful consideration in the specification and quality control of the concrete should be ensured.

1.1.2 Concrete specification

To specify concrete it is necessary to select its characteristic strength together with any limits required on mix proportions, the requirements of fresh concrete and the type of materials that may be used. It is also important to understand the methods of transport, placing and compaction that are to be used as they can change the characteristics and performance of a slab. Figure 1 shows the construction of a typical in-situ concrete industrial hardstanding. It is current practice to specify either a 'designed mix' or a special 'prescribed mix' and where applicable, compliance testing procedures (shown in Fig. 2) should be performed. BS 5328: 1981, *Methods of specifying concrete*,[1.1] defines the two types of mixes as follows.

> A *designed mix* is specified by its required performance in terms of strength grade, subject to any special requirements for materials, minimum or maximum cement content, maximum free water/cement ratio and any other properties required. Strength testing forms an essential part of compliance.

Fig. 1.   Concrete being placed and compacted by a poker vibrator. The concrete is being discharged from a readymixed truck into a bay with timber formwork and steel mesh reinforcement

Fig. 2.   Tests are undertaken on site as the concrete is placed. The technician is undertaking a slump test to measure the workability of the concrete and is making cubes for strength testing

A *prescribed mix* is specified by its constituent materials and the properties or quantities of those constituents to produce a concrete with the required performance. The assessment of the mix proportions forms an essential part of the compliance requirements. Strength testing is not used to assess compliance.

A prescribed mix should be specified only when there is reliable previous evidence or data established from trial mixes that with the materials and workmanship available the concrete produced will have the required strength, durability and other characteristics. This type of mix may be required to produce concrete having particular properties, e.g. to obtain a special finish.

1.1.3 Types of cementitious material

Commonly specified cementitious materials are listed below with their appropriate British Standard references.

BS 12:	1978	*Ordinary and rapid hardening Portland cement*
BS 146:	1973:	*Part 2 Portland blast furnace cement*
BS 1370:	1979	*Low heat Portland cement*
BS 3892:	1982:	*Part 1 Pulverised fuel ash for use as a cementitious component in structural concrete*
BS 4027:	1980	*Sulphate-resisting Portland cement*
BS 6588:	1985	*Portland pulverised fuel ash cement*
BS 6699:	1986	*Ground granulated blast furnace slag (GGBS) for use with Portland cement*

Ordinary or rapid hardening Portland cement is the most common cementitious material used, but other cements or other combinations of Portland cement with ground granulated blast furnace slag (GGBS) and pulverised fuel ash (PFA) may be used provided that satisfactory data on their suitability, such as previous performance tests, are available. It has been suggested[1.2] that the replacement of up to 35% cement by PFA, or 50% GGBS, could be undertaken without an adverse effect on the skid resistance or flexural strength of the concrete slab, provided thorough curing for at least seven days has been performed. In the case of GGBS the flexural strength exhibited by the concrete may be enhanced.

1.1.4 Aggregates

Aggregates should comply with BS 882: 1992, *Aggregates from natural sources for concrete*.[1.3] The following recommendations should be complied with if a durable surface is to be achieved. The physical/mechanical properties defined are determined in accordance with BS 812 and stated in BS 882.

- 10% fines value should not be less than 100 kN.
- Aggregate impact value should be not less than 30%.

- Flakiness index should not exceed 35.
- Drying shrinkage of concrete should be less than 0.065%.

1.1.4.1 Fine aggregate. The fine aggregate is usually naturally occurring sand. Aggregate passing a 5 mm BS 410 test sieve is termed sand. BS 882: 1992: Section 5.2.2 gives the grading limits of sand used for the construction of concrete hardstandings. Sands within grading C or M should be used. Sand may be specified as either uncrushed, crushed or blended. Uncrushed sand results from the natural disintegration of rock, whereas crushed sand is the product of crushing processes of gravel or rock. Blended sand is a controlled mixture of two or more of the types described. Very coarse or very fine gradings as well as gap gradings should not be used as this can often lead to difficulties in finishing or poor durability of the surface. The sand should be free from soft materials, such as soft sandstone, limestone, coal and lignite, and the use of unwashed crushed fines can seriously inhibit the quality of the slab.

1.1.4.2 Coarse aggregate. For most slabs 20 mm maximum size aggregate is suitable, but for the construction of thicker slabs it may be more economical to use aggregates up to 40 mm maximum size. Gradings as defined in BS 882[1.3] should be used. Soft sandstone or soft limestone should be avoided and crushed igneous or crushed flint gravels of angular shapes are preferable. Aggregates other than those stated in BS 812: Part 1 may be used provided satisfactory data on the exhibited properties of concrete made with them are available. Recent research[1.4] has found that the use of angular or crushed aggregate has resulted in an increase in flexural strength of up to 25% as compared with the use of rounded or irregular aggregates.

1.1.5 Admixtures

It is common practice to specify admixtures to aid the workability of fresh concrete without loss of strength or durability. Admixtures are permitted in designed and prescribed mixes and, if specified, they should comply to BS 5075: Part 1[1.5] and Part 3[1.6] as appropriate. The use of an additive will normally be determined by the contractor according to his or her method of construction. If a construction procedure requires a self levelling concrete, no water should be added to the mix on site. In this case the use of a super-plasticizer is common. A super-plasticizer is defined in BS 5075: Part 3[1.6] as an *admixture that, when added to a hydraulic binder concrete, imparts very high workability or allows a large decrease in water content for a given workability*. If two or more admixtures are to be used simultaneously, care should be taken to assess their interaction and to ensure their compatibility.

A common admixture is air entraining agent. In-situ concrete hardstandings in the UK are likely to develop the type of surface deterioration shown in Figs 3 and 4 as a result of the mechanical effects of freezing water. If moisture within the concrete freezes, it expands by 10%

Fig. 3.   An example of concrete damaged by frost. The concrete surface has been lost revealing the aggregate. This usually occurs as a result of the absence of entrained air or poor workmanship

Fig. 4.   This bay is in an advanced state of frost damage. Not only has the surface been lost but also the concrete has begun to erode. The reinforcement may start to corrode as the electrochemical protection offered by the cement is lost

and applies massive forces within the concrete as it seeks the space to accommodate its greater volume. By entraining air in the form of isolated bubbles in the concrete, room for the ice is created so the stresses do not develop. The bubbles also improve the workability of the fresh concrete so, for a given level of compactive effort, a greater density is achieved which inhibits the ingress of moisture. Air is introduced at the volumetric rate of approximately 4.5% (it can be measured by an air meter) by including a foaming agent in the concrete mix.

Particular care should be taken when dealing with ice on concrete hardstandings (Fig. 5). Rock salt is often spread over the surface of the concrete at a rate of 3 g/m^2 to prevent the formation of ice. The salt forms a brine solution with the precipitation and so depresses the freezing point to $-4°\text{C}$ which is usually low enough to prevent the formation of surface ice in the UK. If ice does form on the surface of the concrete, salt should not be used retrospectively. Applying salt to ice melts the ice by taking heat from the underlying concrete very rapidly. The latent heat of melting of ice (i.e. the kinetic energy needed to mobilize the molecules of the water) is massive and one or two cycles of rapid thawing can destroy a hardstanding surface. Furthermore, if vehicles enter an unsalted frozen hardstanding from a pre-salted public highway, brine may fall from the vehicle onto the ice and damage the concrete.

Fig. 5. Many industrial hardstanding operators apply salt to prevent the formation of ice. Providing the salt is spread before ice forms, the concrete suffers no harm. If salt is applied onto ice, the ice melts immediately and in so doing depresses the temperature of the concrete very quickly. The resulting thermal shock may crack the concrete. Salt solution may also be brought to the hardstanding by vehicles which have been on salted public roads — this can also damage the hardstanding concrete

1.1.6 Concrete quality

Unlike most structural applications of concrete, industrial hardstanding design is based on the flexural strength of the concrete. A relationship between the 28 day characteristic compressive strength and the flexural strength of concrete is required. The 28 day characteristic compressive strength in units of N/mm^2 is defined as the grade of the concrete and is prefixed by the letter C. The relationship between flexural strength and the 28 day compressive strength is given in Table 1.1.[1.7]

From Table 1.1 the relevant flexural strength can be determined and factors of safety applied. This results in the design flexural strength used in Chapter 4.

In order to obtain concrete of a particular strength there are a number of mix design limits that must be complied with. Developments in the technology of cement manufacture in recent times have resulted in the achievement of higher strengths from mixes than those achieved previously. Table 1.2 shows mix design guidance[1.8] for various grades of concrete. To achieve a durable surface, concrete with a minimum cement content of 325 kg/m^3 and water/cement ratio not exceeding 0.55 is commonly specified, although stricter limits may be applied if the hardstanding is to be subject to heavy industrial use.

The workability of fresh concrete should be suitable for the conditions of handling and placing, so that after compaction and finishing the concrete surrounds all reinforcement and completely fills its formwork. The workability of the concrete is normally determined by the contractor to suit his method of working and a slump of 50 mm is the usual maximum. If a slump of greater than 50 mm is allowed there is a tendency for the aggregate

Table 1.1. *Relationship between concrete grade, compressive strength and flexural strength*

Concrete grade	C30	C40	C50
Characteristic compressive strength: N/mm^2	30	40	50
Flexural strength: N/mm^2	3.8	4.5	5.1

Table 1.2. *Relationship between free water/cement ratio, cement content and lowest grade of concrete*

Maximum free water/cement ratio	0.65	0.60	0.55	0.50	0.45
Minimum cement content: kg/m^3	275	300	325	350	400
Grade of concrete: N/mm^2	C30	C35	C40	C45	C50

particles to segregate. Where the consistency of the mix is such that the concrete is unable to hold all its water, some is gradually displaced and rises to the surface. Separation of water from a mix in this manner is known as bleeding and can lead to dusting and poor wear resistance properties of the hardened concrete. The slump of a mix should therefore be carefully monitored on site during pouring.

The lower the water/cement ratio the higher is the strength of the hardened concrete (see Table 1.2), but this can lead to a loss in workability of fresh concrete. Although modern equipment is capable of handling less workable mixes, the use of super-plasticizers (see Section 1.1.5) to increase workability is now common. Lowering the free water/ cement ratio in a concrete mix also has the advantage of enabling finishing to begin earlier.

1.2   Subgrade, sub-base and capping
1.2.1   Subgrade
Subgrade is the naturally occurring ground or imported fill at formation level. Homogeneity of the subgrade strength is particularly important and avoiding hard and soft spots is a priority in subgrade preparation. Any imported subgrade fill should be suitable material of such grading that it can be well compacted. Fill containing variable piece sizes often proves difficult to compact, giving rise to settlement and early failure of the hardstanding. On very good quality subgrades, such as firm sandy gravel the sub-base material may be omitted, although contractors often elect to provide a miminal 150 mm to facilitate construction.

1.2.2   Sub-base and capping
The sub-base and the capping comprise the foundation to the industrial hardstanding (see Fig. 6). For most types of subgrade, a sub-base is essential. This layer usually consists of an inert, well graded granular material (see Section 1.2.5) or cement-treated material such as lean concrete or cement-bound granular material. In-situ cement stabilization may prove an economic means of improving a poor subgrade. In the case of wheel and point loading (see Chapter 3), the sub-base assists in reducing the vertical stress transmitted to the subgrade. When the California Bearing Ratio (CBR) of the subgrade is below 5%, it is sometimes the case that part of the sub-base can be replaced with a lower quality material called capping. For example, it is common to limit the sub-base thickness to 150 mm or 225 mm (the greatest thickness which can be installed in one layer) and to provide capping material beneath to provide the requisite support to the slab. This is possible because the upper part of the foundation is subjected to higher levels of stress than the

Fig. 6.   In the upper picture the surface of the sub-base has been rolled to remove any upwards projections which might penetrate the slip membrane. If the sub-base material is open textured, it may be 'blinded' with fine material prior to the installation of the slip membrane. This sub-base material is just at the point where blinding would be beneficial. Blinding reduces the coefficient of friction and thereby improves the probability that cracking will occur at induced joints rather than in mid-bay locations. In the lower picture, a drain has been bedded on the sub-base. Concrete will be placed at each side of the drain to strengthen its walls. A full movement joint will be required between that concrete and the main hardstanding. This type of drain has 'built-in' falls so it can be installed over a horizontal sub-base surface. An outlet beneath one of the drainage units takes surface water to the main drainage

remainder. Section 4.8 gives recommendations for the thickness of sub-base and capping. Capping is sometimes defined as low-cost locally available material which can achieve a CBR of 15% within itself. Materials such as selected hardcore, crushed concrete and crushed rock failing to meet the grading requirements of sub-base materials are commonly used as capping.

1.2.3   Modulus of subgrade reaction

In assessing the stresses induced in a slab under loading, the influence of the subgrade is introduced as if it were an elastic medium with a modulus of subgrade reaction (K). The modulus of subgrade reaction characterizes the deflexion of the ground and/or the foundation under the industrial hardstanding. California Bearing Ratio (CBR) tests and plate bearing tests can be used to establish values (see Section 1.2.4). In many instances subgrades are variable and results obtained from in-situ tests can often show scatter. CBR and plate bearing tests induce a shallow stress bulb and may not reflect the influence of deeper material which might become stressed beneath a loaded slab. Assumed values of K are shown in Table 1.3. Because the stresses in a concrete industrial hardstanding are insensitive to changes in the strength of the supporting material, the values in Table 1.3 may be assumed if no plate bearing test or CBR test results are available.

Chandler and Neal[1.7] suggest that the sub-base can be taken into account by enhancing the effective modulus of subgrade reaction (K) as in Table 1.4.

When a lean concrete sub-base is specified, the value of K of the subgrade material is used to calculate the required thickness of the concrete slab. This calculated thickness is then apportioned between the structural slab thickness (the higher strength concrete) and the lean concrete sub-base thickness. This relationship is shown in Table 1.5 when a C40 concrete is used for the slab and a C20 lean concrete is used for the sub-base. Figure 7 shows an example of hardstanding failure resulting from inadequate design.

Table 1.3.   Assumed modulus of subgrade reaction (K) for typical British soils[1.7]

	Typical soil description	Subgrade classification	Assumed K: N/mm^3
Coarse grained soils	Gravels, sands, clayey or silty gravels/sands	Good	0.054
Fine grained soils	Gravely or sandy silts/clays, clays, silts	Poor, very poor	0.027, 0.013

Table 1.4. *Enhanced value of K when a sub-base is used*

K value of subgrade alone: N/mm^3	Enhanced value of K when used in conjunction with:							
	Granular sub-base of thickness (in mm)				Cement-bound sub-base of thickness (in mm)			
	150	200	250	300	100	150	200	250
0.014	0.018	0.022	0.027	0.033	0.045	0.063	0.081	0.106
0.027	0.034	0.038	0.044	0.051	0.075	0.104	0.137	—
0.054	0.059	0.065	0.072	0.081	0.125	0.175		
0.082	0.089	0.096	0.105	0.114				

Table 1.5. *The modified thickness of a concrete slab with a C20 lean concrete sub-base*

Calculated thickness of slab: mm	Modified thickness of slab (mm) when used in conjunction with lean concrete sub-base of thickness (mm):		
	100	130	150
250	190	180	—
275	215	200	—
300	235	225	210

Fig. 7. *This hardstanding failed after a few years' use as a result of poor subgrade and insufficient slab and sub-base thickness*

1.2.4 Plate bearing and CBR testing

The plate bearing test procedure is to load the ground through a steel disc, usually mounted on the back of a vehicle, and to record load and corresponding deflexion. The value of K is found by dividing the pressure exerted on the plate by the resulting vertical deflexion and is expressed in units of N/mm^3, MN/m^3 or kg/cm^3. K is established by plate bearing tests with a plate loading diameter of 750 mm. A modification is needed if a different plate diameter is used: for a 300 mm diameter plate K is divided by 2.3 and for a 160 mm diameter plate it is divided by 3.8. Alternatively the CBR can be measured and CBR values expressed as percentages are obtained. The CBR of a soil is determined by a penetration test which measures the force required to produce a given penetration in the material. This force is compared with the force required to produce the same penetration in a standard crushed limestone. The result is expressed in percentage terms as a ratio of the two penetration forces. Thus a material with a CBR value of 4% offers 4% of the resistance to penetration as compared to that offered by standard crushed limestone. The laboratory test should be carried out in accordance with BS 1377. Different subgrade materials will have different CBR values, and a conservative value is used for each category of soil. It is unusual for CBR to be measured directly since it can usually be determined with sufficient accuracy from Liquid Limit (LL) and Plasticity Index (PI) values. If the CBR is to be measured directly, it should be done so at the most adverse moisture content which the soil can reasonably be predicted to sustain. BS 1377 includes a 72 hour soaking procedure which will be appropriate in some design situations. Table 1.6[1.9] shows the relationship between CBR and K for a number of common soil types.

The standard method of classifying soils in the US for engineering purposes is the Unified System and this system can also be used to assess CBR and K. The Unified System classifies soils on the basis of grain size

Table 1.6. Modulus of subgrade reaction and CBR values for a number of common subgrade and sub-base materials

	CBR: %	Modulus of subgrade reaction K: N/mm^3
Humus soil or peat	<2	unacceptable
Recent embankment	2	0.01–0.02
Fine or slightly compacted sand	3	0.015–0.03
Well compacted sand	10–25	0.05–0.10
Very well compacted sand	25–50	0.10–0.15
Loam or clay (moist)	3–15	0.03–0.06
Loam or clay (dry)	30–40	0.08–0,10
Clay with sand	30–40	0.08–0.10
Crushed stone with sand	25–50	0.10–0.15
Coarse crushed stone	80–100	0.20–0.25
Well compacted crushed stone	80–100	0.20–0.30

and plasticity. The initial division of soils is based on the separation of coarse (sand) and fine (clay) grained soils and highly organic soils (peat). The distinction between coarse and fine grained is determined by the amount of material retained on a No 200 (75 micron) sieve. Coarse grained soils are subdivided into sands and gravels on the basis of the amount of material retained on a No 4 (6 mm) sieve. Gravels and sands are then classed according to whether fine material is present. Fine grained soils are subdivided into two groups on the basis of LL and PI. Where the soil has been classified in this way, it may be more convenient to use the K values below rather than the ranges in Table 1.6. The classification system subdivides soil types into different groupings according to the following system.

GW Well graded gravels and gravel — sand mixtures, little or no fines, $k > 0.082 N/mm^3$

GP Poorly graded gravels and gravel — sand mixtures, little or no fine, $k > 0.082 N/mm^3$

GM Silty gravels, gravels — sand mixtures, $k = 0.082 N/mm^3$

GC Clayey gravels, gravel — sand — silt mixtures, $k = 0.054 N/mm^3$

SW Well graded sands and gravelly sands, little or no fines, $k = 0.054 N/mm^3$

SP Poorly graded sands and gravelly sands, little or no fines, $k = 0.054 N/mm^3$

SM Silty sands, sand — silt mixtures, $k = 0.054 N/mm^3$

SC Clayey sands, sand — clay mixtures, $k = 0.054 N/mm^3$

ML Inorganic silts, very fine sands, rock flour, silty or fine sands, $k = 0.027 N/mm^3$

CL Inorganic clays of low to medium plasticity, gravely clays, silty clays, lean clays, $k = 0.027 N/mm^3$

OL Organic silts and organic silty clays of low plasticity, $k = 0.027 N/mm^3$

MH Inorganic silts, micaceous or diatomaceous fine sands or silts, plastic silts, $k = 0.027 N/mm^3$

CH inorganic clays of medium to high plasticity, $k = 0.014 N/mm^3$

PT Peat, mud and other highly organic soils – unnacceptable

In the above list G = Gravel, S = Sand, C = Clay, W = Well, P = Poor, M = Medium, H = High plasticity, L = Low plasticity, O = Organic, PT = Peat and K values have been rationalized to four values which are used in design. The Unified System allows soils to be classified from any geographic location into categories to which engineering properties can be assigned, e.g. particle size distribution, LL and PI. The various groupings of this classification system have been devised to correlate in a general way with the engineering behaviour of soils. This procedure provides a useful step in any site or laboratory investigation for geotechnical engineering purposes.

1.2.5 Granular sub-base materials

The information provided in this section is based upon *Specification for Highway Works, Series 800, Road Pavements — Unbound Materials,*[1.10a] henceforth referred to as the DTp specification. The material should comprise an approved durable granular material such as gravel, hard clinker, crushed rock or well burnt colliery shale, blended if necessary with sand or other fine screenings.

Blast furnace slag for use as a sub-base material should comply with BS 1047. Steel slag may be used provided it has been weathered and conforms to the requirements of BS 4987: Part 1. Materials other than slag when placed within 500 mm of cement-bound materials or concrete products should have a water soluble sulphate content not exceeding 1.9 g of sulphate (expressed as weight of SO_3 per litre) when tested in accordance to BS 1377: Part 3.

1.2.5.1 DTp granular sub-base material Type 1. Unless evidence suggests that Type 2 materials will be suitable, all granular sub-bases should be constructed from Type 1 materials which can comprise crushed rock, crushed slag, crushed concrete or well burnt non-plastic shale. The material must lie within the grading envelope of Table 1.7 and not be gap graded. The sub-base material is transported, laid and compacted without drying out or segregation. The material must have a ten per cent fines value of 50 kN or more when tested to BS 812: Part 111 and an Aggregate Crushing Value (ACV) of not less than 30 when tested to BS 812: Part 111. Additionally, the material should have a CBR of 30% or more.

Table 1.7. Grading requirements for granular materials

BS sieve size	Percentage by mass passing	
	Granular sub-base material Type 1	Granular sub-base material Type 2
75 mm	100	100
37.5 mm	85–100	85–100
10 mm	40–70	40–100
5 mm	25–45	25–85
600 micron	8–22	8–45
75 micron	0–10	0–10

1.2.5.2   DTp granular sub-base material Type 2.   Type 2 granular materials are made up of natural sands, gravels, crushed rock, crushed slag, crushed concrete or well burnt non-plastic shale. The DTp specification states that the material must lie within the grading envelope of Table 1.7 and not be gap graded. The material is transported, laid and compacted at a moisture content within the range 1% above and 2% below the optimum moisture content and without drying out or segregation. The material must have a ten per cent fines value of 50 kN or more when tested to BS 812: Part 111. Additionally, the material should have a CBR of 20% or more.

1.2.5.3   Compaction of granular materials.   Unbound material up to 225 mm compacted thickness is spread and compacted in one layer so that after compaction the total thickness is as specified. The minimum compacted thickness should not be less than 110 mm. Where the layers of unbound material are of unequal thickness the lowest layer should be the thickest layer. Compaction of unbound materials is carried out by the methods shown in Table 1.8. The surface of any one layer of material on completion of compaction and immediately before overlaying should be well closed, free from movement under compaction plant and from ridges, cracks, loose material, pot holes, ruts or other defects. All loose, segregated or otherwise defective areas should be removed to the full thickness of the layer, and new material laid and compacted.

1.2.6   Cement-stabilized sub-bases

The information provided in this section is based upon *Specification for Highway Works, Series 1000, Road Pavements — Concrete and Cement Bound Materials.*[1.10b]

1.2.6.1   Constituents.   The cement in cement-bound materials must comply with the materials in Table 1.9 or the combinations in Table 1.10

The maximum proportions of ground granulated blast furnace slag (GGBS) with Portland cement should not be greater than 65% of the total cement content for cement-bound materials. The water content should be the minimum amount required to provide suitable workability to give full compaction and the required density.

1.2.6.2   Cement Bound Material Category 1 (CBM1).   CBM1 is typically made from a material which has a grading finer than the limits in Table 1.11

1.2.6.3   Cement Bound Material Category 2 (CBM2).   CBM2 is typically made from gravel–sand, a washed or processed granular

Table 1.8. Compaction requirements for granular sub-base material Types 1 and 2

Type of compaction plant	Category[b]	Number of passes for layers not exceeding the following compacted thicknesses (in mm)[a]		
		110	150	225
Smooth-wheeled roller (or vibratory roller operating without vibration	Mass per metre width or roll: over 2700 kg up to 5400 kg over 5400 kg	16 8	Unsuitable 16	Unsuitable Unsuitable
Pneumatic-tyred roller[c]	Mass per wheel: over 4000 kg up to 6000 kg over 6000 kg up to 8000 kg over 8000 kg up to 12 000 kg over 12 000 kg	12 12 10 8	Unsuitable Unsuitable 16 12	Unsuitable Unsuitable Unsuitable Unsuitable
Vibratory roller[d]	Mass per metre width of vibrating roll: over 700 kg up to 1300 kg over 1300 kg up to 1800 kg over 1800 kg up to 2300 kg over 2300 kg up to 2900 kg over 2900 kg up to 3600 kg over 3600 kg up to 4300 kg over 4300 kg up to 5000 kg over 5000 kg	16 6 4 3 3 2 2 2	Unsuitable 16 6 5 5 4 4 3	Unsuitable Unsuitable 10 9 8 7 6 5
Vibrating plate compactor[e]	Mass per square metre of base plate: over 1400 kg/m^2–1800 kg/m^2 over 1800 kg/m^2–2100 kg/m^2 over 2100 kg/m^2	 8 5 3	 Unsuitable 8 6	 Unsuitable Unsuitable 10
Vibro-tamper[f]	Mass: over 50 kg up to 65 kg over 65 kg up to 75 kg over 75kg	4 3 2	8 6 4	Unsuitable 10 8
Power rammer[g]	Mass: 100 kg up to 500 kg over 500 kg	5 5	8 8	Unsuitable 12

Notes to Table 1.8

[a] The number of passes is the number of times that each point on the surface of the layer being compacted is traversed by the item of compaction plant in its operating mode (or struck in the case of power rammers).

[b] The compaction plant is categorized in terms of static mass. The mass per metre width of roll is the total mass on the roll divided by the total roll width. Where a smooth-wheeled roller has more than one axle, the category of the machine is determined on the basis of the axle giving the highest value of mass per metre width.

[c] For pneumatic-tyred rollers the mass per wheel is the total mass of the roller divided by the number of wheels. In assessing the number of passes of pneumatic-tyred rollers the effective width is the sum of the widths of the individual wheel tracks together with the sum of the spacings between the wheel tracks providing that each spacing does not exceed 230 mm. Where the spacings exceed 230 mm the effective width is taken as the sum of the widths of the individual wheel tracks only.

[d] Vibratory rollers are self propelled or towed smooth-wheeled rollers having means of applying mechanical vibration to one or more rolls. The requirements for vibratory rollers are based on the use of the lowest gear on a self propelled machine with mechanical transmission and a speed of 1.5–2.5 km/h for a towed machine. Vibratory rollers operating without vibration are classified as smooth-wheeled rollers.

[e] Vibrating-plate compactors are machines having a base plate to which is attached a source of vibration consisting of one or two eccentrically weighed shafts. They normally travel at speeds of less than 1 km/h

[f] Vibro-tampers are machines in which an engine-driven reciprocating mechanism acts on a spring system, through which oscillations are set up in a base plate.

[g] Power rammers are machines which are actuated by explosions in an internal combustion cylinder, each explosion being controlled manually by the operator. One pass of a power rammer is considered to have been made when the compacting shoe has made one strike on the area in question.

Table 1.9.   Cementitious material specifications

Cement	Complying with
Portland cement (PC)	BS 12
Portland blast furnace cement (PBC)	BS 146
Portland pulverized fuel ash cement	BS 6588
Pozzolanic cement (Grades C20 or below)	BS 6610

material, crushed rock, all-in aggregate, blast furnace slag or any combination of these. The constituents of the material must fall within the grading limits shown in Table 1.11. The material must have a ten per cent fines value of 50 kN or more when tested in accordance with BS 812: Part 111 with samples in a soaked condition.

Table 1.10. Cementitious material combination specifications

Combination	Complying with
Portland cement with ground granulated blast furnace slag	BS 12
Portland cement with pulverized fuel ash for use as a cementious component	BS 3892: Part 1
Portland cement with micro-silica having a current BBA certificate	BS 12

Table 1.11. Grading of aggregate materials used in the four categories of cement bound materials

BS sieve size	Percentage by mass passing nominal maximum size:			
	–	–	40 mm	20 mm
	CBM1	CBM2	CBM3 & CBM4	
50 mm	100	100	100	–
37.5 mm	95	95–100	95–100	100
20 mm	45	45-100	45–80	95–100
10 mm	35	35-100	N/A	N/A
5 mm	25	25-100	25-50	35–55
2.36 mm	N/A	15–90	N/A	N/A
600 micron	8	8–65	8–30	10–35
300 micron	5	5–40	0–8	8–8
75 micron	0	0–10	0–5	0–5

1.2.6.4   Cement Bound Material Category 3 (CBM3).   CBM3 is made from natural aggregate material complying with BS 882.

1.2.6.5   Cement Bound Material Category 4 (CBM4).   CBM4 is made from natural aggregate material complying with BS 882.

If blast furnace slag aggregate is to be used, it must comply with BS 1047: 1983. Cement for use in all cement-bound material and aggregate for use in CBM3 and CBM4 should be kept dry and used in the order in which it is delivered to the site. Different types of cementitious material must be stored separately.

1.2.6.6   Drylean concrete.   Drylean concrete is a lean concrete with a low water content. The maximum aggregate to cement ratio is 15 to 1. The water content should be between 5 and 7% by weight of dry

Table 1.12.   Batching and mixing of cement-bound materials

	Site requirements				Specimen requirements	
Category	Mixing plant	Methods of batching	Moisture content	Minimum compaction	Minimum 7 day cube compressive strength: N/mm^2	
					Average	Individual
CBM1	Mix-in-place or mix-in-plant	Volume or mass	To suit requirements for strength, surface level, regularity and finish	95% of cube density	4.5	2.5
CBM2	"	"	"	"	7.0	4.5
CBM3	Mix-in-plant	Mass	"	"	10.0	6.5
CMB4	"	"	"	"	15.0	10.0
Drylean Concrete	"	"	Between 5% & 7% of dry weight	Maximum possible	15.0 (Maximum) (No single cube below 12)	

materials, the final value being selected to give the maximum dry density. The material should be rolled to give the maximum possible density.

1.2.6.7 Batching and mixing.   Cement-bound materials should be made and constructed as detailed in Table 1.12.

Batching and mixing should be carried out in the appropriate manner described in Table 1.12. Where the mix-in-plant method is used and materials are batched by mass, materials should be batched and mixed in compliance with BS 5328: Part 3.

1.2.6.8 Transporting.   Plant-mixed cement-bound material when mixed should be removed from the mixer immediately and transported directly to the point in consideration.

1.2.6.9 Laying.   All cement-bound material should be placed and spread evenly in such a manner as to prevent segregation and drying.

Spreading the material is undertaken concurrently with placing or without delay. Base cement-bound material is often spread using a paving machine or a spreader box and operated with a mechanism which levels off the cement-bound material to an even depth. Cement-bound material is always spread in one layer so that after compaction, the total thickness is as specified. Compaction is carried out immediately or within 2 h of the addition of the cement. The surface of any layer of cement-bound material on completion of compaction and immediately before overlaying, should be well closed, free from movement under compaction plant and from ridges, cracks, loose material, pot holes, ruts or other defects.

1.2.6.10 Compaction.   Compaction should be carried out immediately after the cement-bound material has been spread and in such a manner to prevent segregation Compaction must be completed within 2 h of the addition of the cement. The surface of any one layer of cement-bound material on completion of compaction and before overlaying should be well closed, free from movement under compaction plant and from ridges, cracks, loose material, pot holes, ruts or other defects.

1.2.6.11 Curing.   Immediately on completion of compaction, the surface of the cement-bound sub-base is cured for a minimum period of seven days.

1.2.7 Settlement

A site investigation can provide the necessary information to enable an estimate of long-term settlement to be made. Where slabs are supported on subgrade such as organic soils, heavy clays and loose sands, or where land has been reclaimed, anticipated long-term settlements may be significant. Plate bearing tests, as described in Section 1.2.4, enable long-term settlements to be predicted. Soil stabilization, drainage or compaction, or the use of piled foundations may be used to reduce or eliminate settlement.

1.3 Slip membranes

A slip membrane consists of polyethylene sheeting laid beneath the concrete slab with overlaps of at least 200 mm and is usually placed immediately prior to concrete pouring. Wrinkles and folds, as shown in Fig. 8, should be completely removed since they can result in weakening of the slab in later life as they may form crack inducers. It is advisable to anchor the polyethylene sheeting with small heaps of concrete, especially on the overlaps. A minimum of 125 micron (500 gauge) polyethylene sheeting should be used and 250 micron (1000 gauge) or 300 micron (1200 gauge) sheets are common.

Fig. 8. Slip membrane in place ready for the reinforcement to be fixed. The water lying on the surface will be pumped away prior to placing the concrete. Care has to be taken to avoid ripples or creases, especially during steel fixing

A slip membrane is used to reduce friction between a concrete slab and its sub-base. The coefficient of friction with the use of a membrane is in the region of 0.2, compared with values of up to 0.7 when the concrete slab and sub-base are in direct contact. Prevention of loss of moisture and fines from the fresh concrete into the sub-base does occur, although a slip membrane is not intended or required to serve as a damp-proof membrane.

If an impermeable membrane is used then drying can take place only from the upper surface of the slab which may result in curling (see Chapter 5). The perforation of the slip membrane, or complete omission with the use of a blinding material, may therefore need to be considered although this would result in an increased loss of water and fines from the underside of the slab.

1.4   Welded steel wire fabric (mesh)

Fabric should comply with the requirements of BS 4483: 1985.[1.11] Information has been taken from BS 4483 in the preparation of this section.

1.4.1   Introduction

Steel wire fabric comprises an orthogonal arrangement of longitudinal wires and cross wires welded together at some or all of the cross-over points in a shear-resistant manner. The fabric is usually manufactured by machine with the intersection joints formed by electrical resistance

Fig. 9.   Steel mesh delivered to site is identified by the label. In this case, A393 indicates square mesh with 393 mm²/m of steel in each orthogonal direction. All sheets are of dimensions 2.4 m × 4.8 m and can be cut to avoid obstructions. Surface corrosion is no problem and can assist in creating a bond between the steel and the concrete

welding. Butt-welded wires are sometimes used. The shearing load required to produce failure of a welded intersection should not be less than $0.25Af_y$, where A is the nominal cross-sectional area of the smaller wire at the welded intersection and f_y is the wire's characteristic yield strength. The mesh is usually supplied in bundles bound together in a flat, rolled or folded form (Fig. 9).

1.4.2   Quality control

Manufacturers specify wire of grade 460 complying with the relevant British Standards (BS 4449, BS 4461, BS 4482) to produce the fabric. The number of broken welds must not exceed 4% of the total and must not exceed half the number of cross-welded joints along any one wire.

1.4.3   Dimensioning

Steel wire fabric is available in the wire diameter and spacing arrangements shown in Table 1.13; an example is shown in Fig. 10.

Table 1.13.   Preferred range of designated fabric types and stock sheet size[1.11]

Fabric reference[a]	Longitudinal wires			Cross wires			
	Nominal size: mm	Pitch: mm[b]	Area: mm²/m	Nominal wire size: mm	Pitch: mm	Area: mm²/m	Mass: kg/m²
Square mesh							
A393	10	200	393	10	200	393	6.16
A252	8	200	252	8	200	252	3.95
A193	7	200	193	7	200	193	3.02
A142	6	200	142	6	200	142	2.22
A98	5	200	98	5	200	98	1.54
Structural mesh							
B1131	12	100	1131	8	200	252	10.9
B785	10	100	785	8	200	252	8.14
B503	8	100	503	8	200	252	5.93
B385	7	100	385	7	200	193	4.53
B283	6	100	283	7	200	193	3.73
B196	5	100	196	7	200	193	3.05
Long mesh							
C785	10	100	785	6	400	70.8	6.72
C636	9	100	636	6	400	70.8	5.55
C503	8	100	503	5	400	49	4.34
C385	7	100	385	5	400	49	3.41
C283	6	100	283	5	400	49	2.61
Wrapping mesh[c]							
D98	5	200	98	5	200	98	1.54
D49	2.5	100	49	2.5	100	49	0.77
Stock sheet size[d]	Longitudinal wires			Cross wires			Sheet area
	Length 4.8 m			Width 2.4 m[e]			11.52 m²

[a] When specifying a steel wire, fabric reference codes should be used. Reference letters A, B, C and D represent square, structural, long and wrapping mesh respectively. The numbers in the reference represent the area of steel of the longitudinal wires per metre width of fabric.

[b] Pitch is defined as the centre to centre spacing of wires in a sheet of fabric.

[c] Wrapping mesh: wire usually of grade 250 for use in wrapping fabric.

[d] Stock sheet size: fabric types A and B are delivered in standard sheets of 4.8 × 2.4 m², or in scheduled size sheets. Fabric type C is available in sheets or rolls.

[e] When specifying fabric sheets the width is the overall dimension measured in the direction of the cross wires.

Fig. 10.   The mesh reinforcement has been fixed onto supports over the slip membrane. Neighbouring sheets overlap by 200 mm. In this case, the mesh is located near the upper surface of the slab in order to reduce the possibility of surface cracking

1.4.4   General

Steel wire fabric is assumed to carry the tensile force developed in the concrete owing to the contraction of the slab due to shrinkage and expansion. Consequently mesh allows greater joint spacings. Fabric is often used in long strip hardstanding construction as it can be placed conveniently without the need for cutting. Where necessary, fabric sheets should be lapped at their edges and ends by 450 mm. Overlapping can result in an unacceptable build up of thickness of reinforcement. When using wire-guided vehicles, interference with control signals needs to be considered, i.e. careful placing at a specific depth within the slab may be needed. Figures 11 to 14 illustrate the use of welded steel wire fabric.

1.5   Fibres as a reinforcing material

During recent years hardstanding construction methods involving the addition of polypropylene or steel fibres into a concrete mix have become common in the UK. For centuries, man has attempted to reinforce construction mortars and concretes with various types of fibres. Pharaoh ordered his foremen to cease supplying reinforcing straw to Israelite brick makers (Exodus 5[7]) and later the Romans used hair fibres in structural mortars.

Fig. 11.   The reinforcement supporting structure has to be able to hold the mesh reinforcement in place during concreting. The concreting gang may stand on the mesh and the poker vibrator will attempt to disturb the arrangement

Fig. 12.   This bay is ready for concreting. The A393 mesh is sufficient to allow transverse joints to be spaced at 40 m

With the advent of fast-track systems in the construction industry, concrete hardstanding installation has had to meet quicker construction programmes. With the use of laser guided screeding machines, fibres are often specified instead of conventional mesh because of the inconvenience in positioning individual mats of mesh immediately in front of the laser screeding machine as the work progresses. Laser

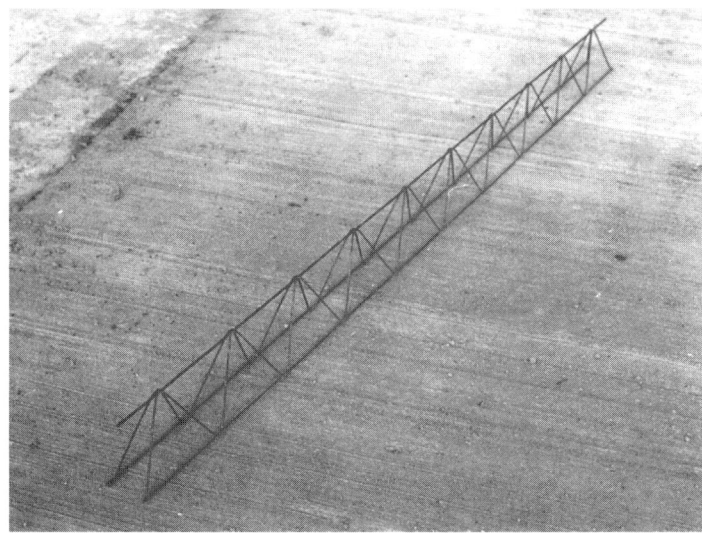

Fig. 13.   The mesh reinforcement support allows the reinforcement to be fixed at the specified height without risking puncturing the slip membrane

Fig. 14.   In this tied joint, the main A393 mesh near the upper surface is omitted and a length of the same mesh is provided at mid-slab depth. This ensures that shrinkage crack control is achieved

screeding machines cannot construct conventional long strip mesh reinforced hardstandings efficiently since both the mesh and the formwork impede the machine. As a consequence, plain concrete or fibre-based concrete is often specified for laser screeded hardstandings.

1.5.1   Steel wire fibres

1.5.1.1   Introduction.   Steel fibres may be used in place of mesh reinforcement. The stresses occurring in an industrial hardstanding are complex and, depending on the type of load, tensile stresses can occur at the top and at the bottom of the slab. There are, in addition, stresses that are difficult to quantify, arising from a number of causes such as sharp turns from handling equipment, shrinkage and thermal effects and impact loads. The addition of steel wire fibres to a concrete slab results in a homogenously reinforced slab achieving a considerable increase in flexural strength and enhanced resistance to shock and fatigue.

1.5.1.2   Concrete composition and quality.   In order to obtain steel fibre reinforced concrete that is easy to pump and to work, with minimum shrinkage, a steel wire fibre manufacturer[1.9] specifies the following.

- Quantity of cement (commonly Ordinary Portland Cement (OPC)) should be between 320 and 350 kg/m^3.
- 750 to 850 kg/m^3 good quality zero to 4 mm well graded sharp sand should be used.
- A continuous aggregate grading with a maximum size of 28 mm for rounded gravel and 32 mm for crushed stone should be used. Limit the fraction larger than 14 mm to 15–20%.
- Characteristic compressive strength of at least 25 N/mm^2 should be used.
- Water/cement ratio should be about 0.50, and should not exceed 0.55.
- The use of a super-plasticizer is permitted to obtain the necessary workability.
- Admixtures of chloride or chloride-containing concrete additives are not permitted.

1.5.1.3   Addition and mixing.   The recommended dosage rate of steel fibres is between 20 and 40 kg/m^3. The greater the dosage rate the greater is the flexural strength of the slab for a particular grade of concrete. Fibres can be added at the mixing plant or on site directly into the mixing truck. At the mixing plant the steel fibres are usually added into the mixer at the same time as the aggregates. On site the concrete must first achieve the correct workability by the addition of a super-plasticizer before the fibres are added (see Section 1.1.5). The fibres should then be added at the manufacturer's specified rate resulting in a uniform distribution. For example, one manufacturer[1.9] recommends addition at a rate of two 30 kg bags per minute with the truck rotating at full mixing speed and mixing continuing for a further two minutes after the addition of the full dose.

Visual inspection during pouring is necessary to check fibre distribution is satisfactory. All fibre bundles must separate into individual fibres, otherwise mixing is insufficient.

1.5.1.4 Placing, curing and finishing. When placing an industrial hardstanding slab the concrete should be compacted as effectively as possible. Conventional means of tamping or vibration can be used. The usual techniques of floating and trowelling can be applied for finishing. Brushing of the concrete surface is usually undertaken. Immediately after finishing, a curing compound should be applied to combat rapid drying, forming an unbroken film on the surface of the concrete. A second curing layer may be applied if environmental conditions might cause rapid drying. Thin hardstandings, with thicknesses of 120 mm or less, should be provided with a double curing layer to prevent the risk of curling at the edges resulting from overfast drying. Plastic sheets must not be applied if there is a risk that the temperature will become too high and result in the concrete setting too quickly.

1.5.1.5 Types of steel fibre available. The most commonly used steel fibre is the 60 mm long hooked fibre. Hooked fibres are usually glued together (collated) with a special water-soluble glue to form fibre plates which readily disperse in the concrete mixer. The hooks help to ensure optimum fibre anchorage (or adhesion) in the hardened concrete. Enhanced adhesion can be achieved by either anchorage points at the ends of fibres (e.g. a pedal or hook) or, in the case of a crimped fibre, by adhesion along the whole length of the fibre. It is usual to consider only fibres with enhanced adhesion for reinforcement in concrete hardstandings. Figure 15 illustrates the range of steel fibre types.

Breaking or premature deformation of the fibres is prevented by the very high tensile strength of the drawn wire (usually greater than 1100 N/mm^2). The aspect ratio, which is the fibre length to fibre diameter ratio, is also an important factor in fibre specification with common values of 60 and 75.

1.5.1.6 Controlling cracking. Steel wire fibres effectively limit the extension of micro-cracks always present in concrete (Fig. 16). In concrete without fibres, tension cannot be transmitted across the crack, i.e. once the tensile capacity of the plain concrete is exceeded, the micro-crack will extend rapidly resulting in brittle failure. The action of steel wire fibres in a concrete slab is to reduce the concentration of stresses near the micro-cracks by:

(*a*) fibres bridging the crack and therefore transmitting some of the load across the crack;

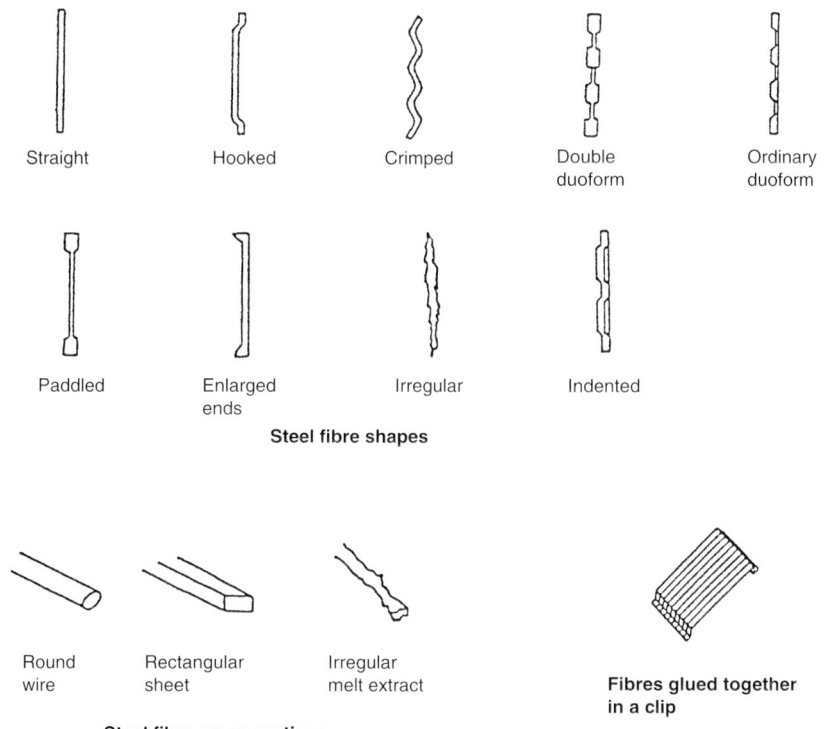

Steel fibre shapes

Steel fibre cross sections

Fig. 15.   *Different steel fibre types*

(*b*) fibres near the crack tip resisting more load owing to their higher modulus of elasticity compared to that of the surrounding concrete.

A crack is formed where the ultimate stress in the slab is exceeded locally. Steel fibres cause the crack to behave as a hinge, resulting in a redistribution of stresses. Unlike a broken zone in a brittle material, this hinge can still resist stresses (depending on the type and dosage used) and thus increases the load bearing capacity of the hardstanding. Chapter 5 shows the results of a research programme on shrinkage cracking of steel fibre reinforced concrete. Steel wire fibres should not be specified to prevent micro-cracking. Micro-cracking in steel fibre reinforced concrete will occur at a similar rate to that expected in plain concrete. Steel fibres prevent micro-cracks from developing into macro-cracks.

1.5.1.7 Flexural strength properties. The primary function of introducing steel fibres into a concrete mix is to increase the load capacity of the slab. Unlike most structural applications of concrete, when

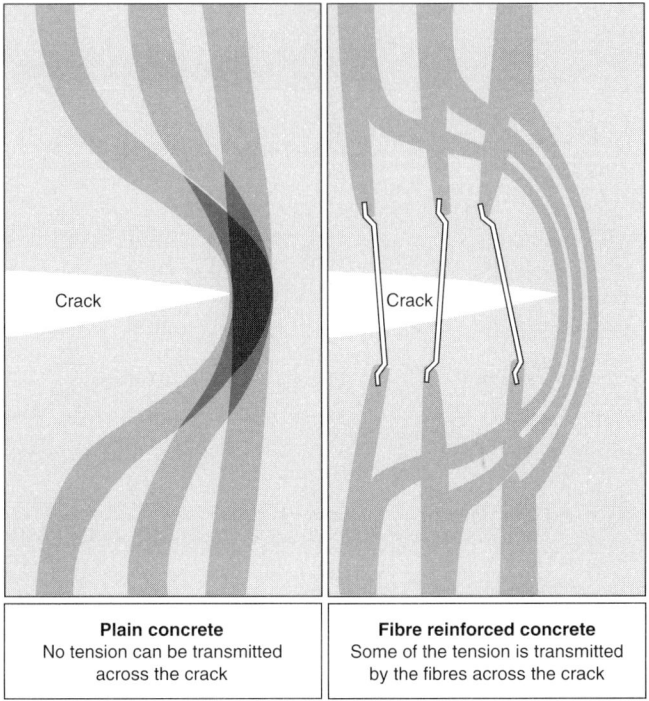

Plain concrete	**Fibre reinforced concrete**
No tension can be transmitted	Some of the tension is transmitted
across the crack	by the fibres across the crack

Fig. 16.   Stress lines in concrete under tension (Tatnall and Kuitenbrouer)[1.12]

designing a hardstanding, we rely upon the flexural strength or modulus of rupture of the concrete, to which we assign a design value. In the US, flexural strength is defined as the stress corresponding to the occurrence of the first crack in the test specimen. This is the point at which the load–deflection curve deviates from a straight line relationship (when the deflexion is δ) as can be seen in Fig. 17. Flexural strength is calculated from the load at first crack and the dimensions of the test specimen.

In the Japanese standard, flexural strength is defined in terms of the maximum load and specimen dimensions as the modulus of rupture. This can be seen in Fig. 18. The flexural strength of concrete is determined from loading tests on concrete prisms (usually $150\,\text{mm} \times 150\,\text{mm} \times 450\,\text{mm}$) at 28 days.[1.13] The test is run at a constant deflection rate of $0.5\,\text{mm/min}$. The actual deflection is recorded as a function of load. The test is continued until the deflection is at least $3\,\text{mm}$ (1/150th of the span). The surface below the curve up to $3\,\text{mm}$ deflection is the flexural toughness D_b, expressed in N/mm. The flexural toughness factor or equivalent flexural strength f_e is defined as:

$$f_e = D_b l / (dbh_t^2) = 1/(150) 2 D_b \qquad (1.1)$$

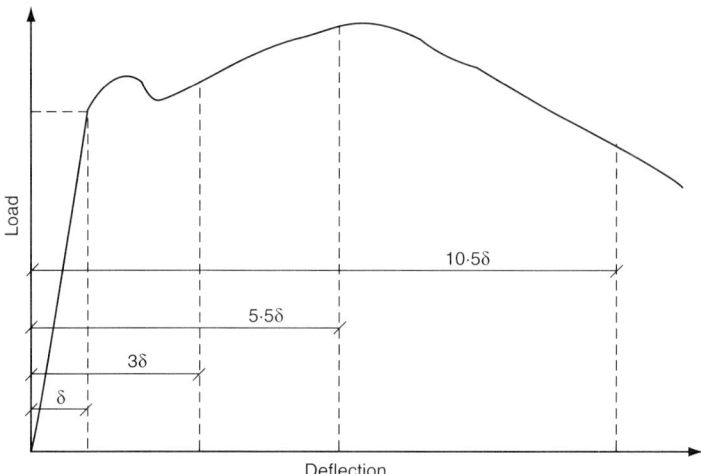

Fig. 17. United States test method. Flexural strength is defined as a deflexion of δ. Often the ultimate flexural strength occurs at a deflexion of 5.5δ

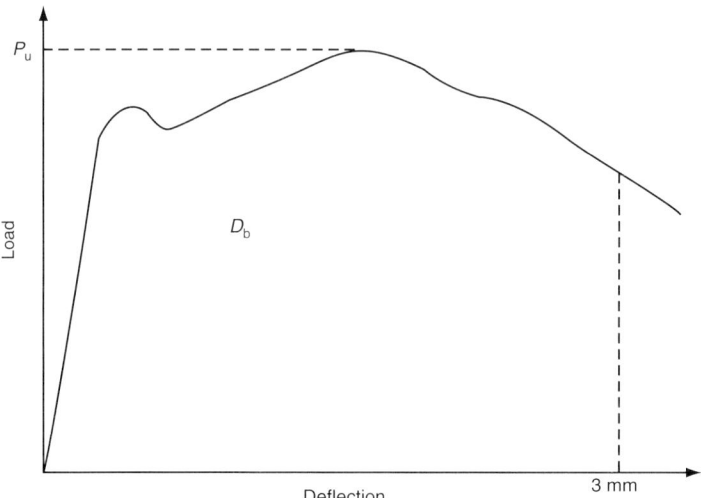

Fig. 18. Japanese test method (P_u = ultimate load)

where D_b is the flexural toughness, I is the second moment of area, d is the deflection of the slab, b is the distance of the applied load from support and h_t is the height of slab (all in millimetres).

The equivalent flexural strength is the representative value for the reinforcing effect of steel fibres. In Japan, this test is already standard, while in the Netherlands it is included in CUR Recommendation $10^{1.14}$ as a basis for determining the design value of Steel Fibre Reinforced Concrete (SFRC). The Dutch design method assumes the mean value of the equivalent flexural strength to be the flexural strength design value f_f.

A series of four point bending tests has been conducted on specimens of dimensions 150 mm × 150 mm × 450 mm with different steel fibre dosages at Newcastle University. Figure 19 illustrates the test used to ensure that the bending moment remains constant over the central third of the span. For each specimen, two load cycles were applied to produce stress/strain relationships similar to that illustrated in Fig. 20. During the first load cycle, the specimen attains a peak load when the concrete first cracks. The specimen is strained to a further degree by a lesser load. The load is removed and a second cycle is applied onto the cracked specimen. A second stress level is determined which represents the effective post-crack flexural strength of the steel fibre reinforced concrete. In fact, the forces are being transmitted through the fibres rather than through the concrete.

Figure 21 shows the effective failure stresses for the first load cycle and the second load cycle for 11 different levels of fibre dosage and for no fibres. Some authorities use the second cycle values in their design: in this book, the first cycle values are used to determine the corresponding characteristic strengths. These values are shown in Table 1.14.

A manufacturer of anchored steel fibres commissioned the TNO in, Delft[1.15] to undertake flexural strength tests using fibres embedded in C30

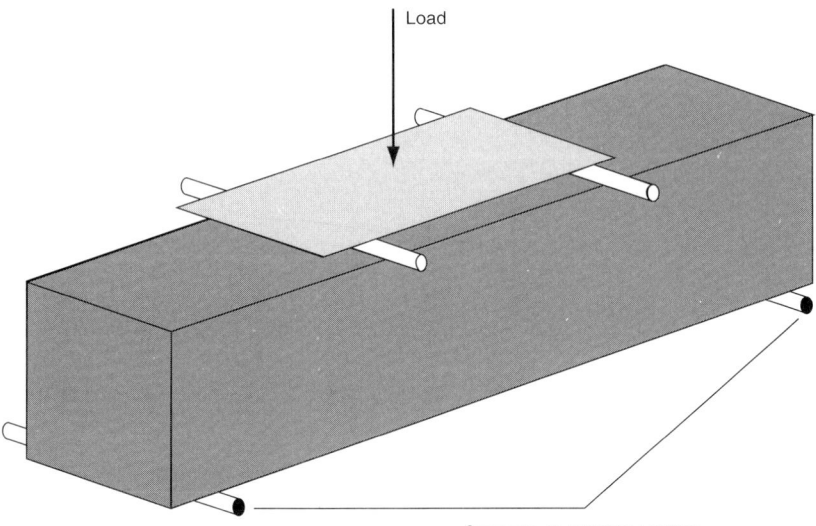

Supports at 450 mm centres

Fig. 19.   Four point bending test arrangement. The central third of the specimen is subjected to a constant bending moment with zero shear force

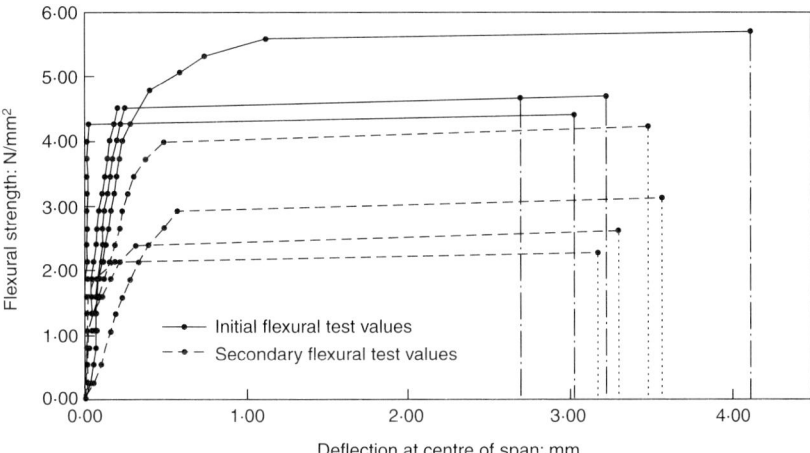

Fig. 20.   Typical stress/strain result from the Newcastle University tests on steel fibre reinforced concrete (50 kg/m³ fibre content)

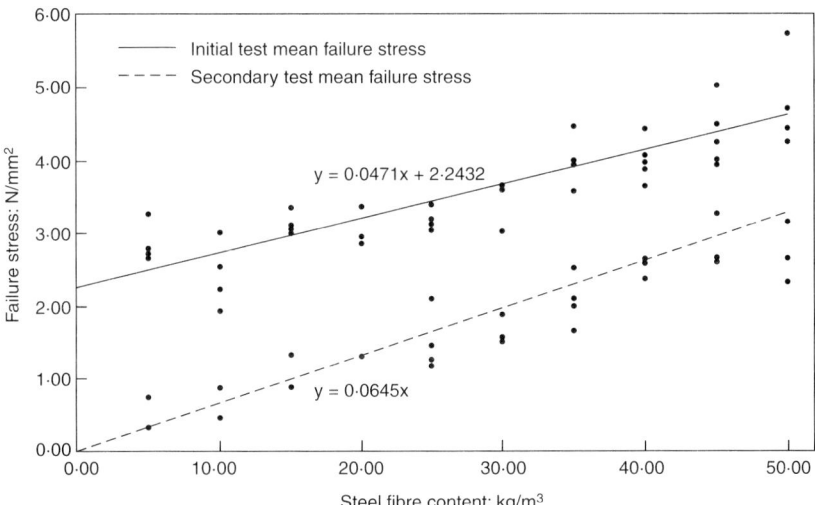

Fig. 21.   Summary of Newcastle University flexural strength test results. The upper line is for the first load application which causes the specimens to crack. The lower line applies to the second load cycle when the concrete is in its post-crack condition

concrete. These tests have resulted in values of mean flexural strength of up to 4.2 N/mm², depending on dosage, type and size of fibre. Partly from these results and partly from work undertaken at the UK Cement and Concrete Association the flexural strength values shown in Table 1.15 have been developed. The values are repeated in Table 4.1. The values in

Table 1.14.   Summary of flexural strength results from Newcastle University tests on steel fibe reinforced concrete

Fibre dosage: kg/m^3	Peak load stress: N/mm^2	Characteristic strength: N/mm^2	Second load cycle stress: N/mm^2
0	2.2	1.5	0
5	2.5	1.8	0.3
10	2.7	2.0	0.7
15	2.9	2.2	1.0
20	3.2	2.5	1.3
25	3.4	2.7	1.7
30	3.7	3.0	1.9
35	3.9	3.2	2.2
40	4.1	3.4	2.6
45	4.4	3.7	2.9
50	4.6	3.9	3.2

Table 1.15.   Concrete mean flexural strengths with steel fibres present

Concrete grade and dosage	Flexural strength: N/mm^2
Plain C30 concrete	2.0
20 kg/m^3 steel fibre C30 concrete	2.8
30 kg/m^3 steel fibre C30 concrete	3.2
40 kg/m^3 steel fibre C30 concrete	3.8
Plain C40 concrete	2.4
20 kg/m^3 steel fibre C40 concrete	3.1
30 kg/m^3 steel fibre C40 concrete	3.6
40 kg/m^3 steel fibre C40 concrete	4.2

Table 1.15 apply to anchored bright wire steel fibres of length 60 mm and wire diameter 1.0 mm.

1.5.1.8   Post-cracking behaviour.   The addition of steel wire fibres to a concrete slab ensures that it has load bearing capacity following the development of the first cracks. Laboratory tests have resulted in the following theory of the behaviour of steel fibre reinforced slabs. Before the first peak load, a concrete slab exhibits elastic behaviour with a modulus of elasticity similar to that of non-fibrous concrete. Following the first peak load and with increasing slab deflection, there is a redistribution of the bending moments which leads to higher ultimate load capacities and thus enhanced performance.

Calculation methods used to determine the thickness of concrete industrial hardstanding slabs are based upon elastic theory and do not take into account the specific properties of SFRC. This has led to some debate concerning the way in which SFRC hardstandings should be designed. The Netherlands CUR Commission has suggested the adoption of a lower elastic modulus for SFRC to account for the redistribution of stresses and has suggested the use of Japanese flexural strength specifications to account for the additional toughness inherent in steel fibre reinforced concrete. Chapter 5 shows the results and conclusions from research programmes studying the post-cracking performance of SFRC.

1.5.1.9   Resistance to impact, fatigue and corrosion.   The increased resistance to fatigue provided by SFRC is of particular importance to hardstandings subject to heavy and intensive traffic. Such hardstandings must be able to resist the frequent and sudden heavy loads common in industrial areas and therefore the increased impact resistance gained by the use of fibres is an important attribute. Spalling of concrete resulting from the corrosion of steel reinforcements is greatly inhibited when SFRC is specified because of the smaller diameter of the steel fibres.

1.5.1.10   Economy.   Steel wire manufacturers claim that substantial labour and materials savings can be achieved by the specification of steel wire fibres. Some of the economic advantages are:

(i)   elimination of labour needed for cutting and fixing traditional mesh reinforcement;
(ii) quicker levelling of the hardstanding owing to the absence of top reinforcement;
(iii) reduction in slab thickness compared with hardstandings designed with plain and mesh reinforced concrete owing to increased flexural strengths;
(iv) greater joint spacing.

The Author has found that suppliers of SFRC increase the price of the concrete by £UK1/m^3 for each kilogram of steel fibres added.

1.5.1.11   Specification.   For any SFRC application it is recommended that the following be included in the specification.[1.14]

- A description of the desired sub-base.
- The required strength class of the concrete.
- The required rate of consistency of the SFRC.
- The method of compaction.
- Usage of a plasticizer if required and which type.
- The method of checking homogeneity of the mix.

1.5.2   Polypropylene fibres

1.5.2.1   Introduction.   Polypropylene fibres for concrete can be in fibrillated or monofilament form. They are manufactured in a continuous process by extrusion of polypropylene homopolymer resin. They are usually coated to improve wetting and dispersibility within the cement paste and to increase the extent of contact and bond between the fibres and the concrete matrix in the hardened state. Polypropylene fibres are not a substitute for conventional structural reinforcement or normal good curing procedures. The design of polypropylene fibre based concrete hardstandings proceeds as for unreinforced hardstandings. The main purpose of polypropylene fibres is to provide crack control by distributing and absorbing tensile stresses which may occur as a result of shrinkage and temperature movements, particularly in the early life of the slab when the concrete has yet to reach sufficient tensile strength. They do not eliminate cracks and are not considered to contribute to the strength of the slab.

1.5.2.2   Monofilament fibres.   Monofilament fibres are manufactured from extruded sheet/film material which is subject to molecular alignment, coated and cut to the appropriate length. This type of fibre is usually much finer than the fibrillated fibre and the properties of a concrete resulting from the addition of monofilament fibres depend on the large number of fibres present. A smoother surface finish may be achieved from the use of the monofilament fibre as opposed to the fibrillated type. Monofilaments do not provide any mechanical bond to the cement paste, but rely on their greater number per metre cube of concrete and their chemical bond in order to achieve their proven qualities in the plastic and hardened state.

1.5.2.3   Fibrillated fibres.   Fibrillated fibres are manufactured from extruded sheet/film material which is subject to molecular alignment, fibrillated, coated and cut to the appropriate length. Clustering of fibres is overcome by the mixing of aggregates in the concrete mix. Basic properties of fibrillated fibre[1.16] are:

- density $= 900\,\text{kg/m}^3$
- tensile strength range $= 560 - 770\,\text{N/mm}^2$
- elastic modulus $= 3.5\text{k N/mm}^2$
- melt point $= 160 - 170°\text{C}$.

Fibrillated fibres have a rough surface texture which gives each fibre a high degree of mechanical bond to the concrete. Monofilament fibres achieve enhanced plastic shrinkage control and trowel workability, while fibrillated fibres impart a higher degree of abrasion resistance to the resulting concrete.

1.5.2.4 Addition and mixing.   The addition of polypropylene fibres is at a recommended dosage of $0.90\,\text{kg/m}^3$ (0.1% by volume or one litre per cubic metre). They are compatible with all cementitious products and admixtures and generally require no change in mix design or water/ cement ratio. The fibres may be added at either a conventional batching/ mixing plant or by hand to the ready mix truck on site. An even distribution throughout the concrete can be achieved in a 6m^3 truck mixer in five minutes at full mixing speed.

1.5.2.5 Placing, curing and finishing.   Concrete mixes containing polypropylene fibres can be transported by normal methods and flow easily from the hopper outlet. No special precautions are necessary when pouring, and fibre-dosed concrete will flow around an obstruction (such as reinforcement) in the same manner as a conventional concrete mix of similar proportions. Conventional means of tamping or vibration to provide the necessary compaction can be used.

Curing procedures similar to those specified for conventional concrete should be strictly undertaken. If steam curing at a temperature in excess of 140°C is to be used polypropylene fibres should not be used. The fibres do not affect the hydration rate or stiffening time of the concrete.

Placed fibre-dosed mixes may be float and brush finished using all normal hand or power tools. Occasional fibres protruding through the surface will quickly wear away. Workmanship should comply with the relevant requirements of BS 8000 Part 2: Sections 2.1 and 2.2.[1.17]

1.5.2.6 Controlling plastic shrinkage cracks in concrete.   Plastic cracking may occur in the plastic concrete as a result of drying shrinkage (see Fig. 22). Plastic cracks are formed within the first 24 h after the concrete has been placed when the evaporation rate is high and the surface

Fig. 22. These are diagonal plastic cracks formed during the concrete setting process as a result of inadequate curing. The water evaporated from the concrete too quickly so wet concrete developed tension and eventually separated along diagonal lines. The cracks have been filled with resin and sealed. Polypropylene fibres can inhibit the formation of plastic cracks

of the concrete dries out rapidly. Plastic shrinkage cracks generally pass through the entire slab and form weakness, permanently lowering the integrity of the slab before the concrete has had the opportunity to gain its design strength. Plastic cracks may occur through the whole depth of a slab and cannot be remedied by surface treatment.

Polypropylene fibres inhibit plastic cracking by holding water at or near the surface of the concrete, delaying evaporation and increasing cement hydration. Therefore bleeding is inhibited. As concrete hardens and shrinks, micro-cracks develop. When the micro-cracks intersect a fibre strand, they are blocked and prevented from developing into macro-cracks and hence plastic cracking. This reduction of micro-cracks in the plastic state enables the concrete to better develop its optimum integrity. A number of research programmes studying the effect on plastic shrinkage cracking of concrete with the use of polypropylene fibres are described in Chapter 5.

1.5.2.7 Effect on workability. Polypropylene fibres act mechanically — the cohesive effect is largely due to surface tension and breaks down under vibration and compaction. The slump of a fibre-dosed concrete will be lower as a result of the thixotropic effect caused by the fibres but the mobility or placeability of the concrete is generally unaffected. Water should not be added to compensate for this thixotropic effect. Vebe and compaction factor tests are not significantly affected by the addition of polypropylene fibres.

The improved cohesiveness also proves to be beneficial in pumped concrete owing to the reduction in rebound when placing.

1.5.2.8 Polypropylene fibre reinforced concrete. The use of polypropylene fibres in the construction of a concrete hardstanding is not considered to contribute to the strength of the slab. The addition of polypropylene fibres at the usual recommended amount ($0.9 \, kg/m^3$) will not significantly affect the ultimate compressive, tensile or flexural strength of the concrete matrix. Before ultimate stress is reached the performance of a fibre-enhanced concrete is improved in a number of ways. These improvements are due to concrete being an inherently variable material with a wide range of stress concentrations and the addition of fibres favourably reduces this variability. If a fibre is aligned across a crack there is a small increase in stress required for crack propagation to occur.

1.5.2.9 Strength characteristics. Tests carried out by a manufacturer of polypropylene fibres[1.16] revealed the change in strength characteristics shown in Table 1.16.

Table 1.16. Test results comparing the strength of polypropylene fibre-dosed concrete and conventional plain concrete

	Strength of fibre reinforced concrete: N/mm^2	Strength of unreinforced concrete: N/mm^2
Compressive strength (equivalent cube method)		
1 day	16.5	16.0
3 days	28.5	24.5
7 days	34.0	35.0
28 days	43.5	39.5
Cube compressive strength		
1 day	16.0	14.5
3 days	28.0	27.5
7 days	34.0	36.0
28 days	48.5	44.5
Flexural strength		
1 day	2.3	2.1
3 days	4.0	3.7
7 days	4.2	4.8
28 days	4.6	6.2

Compressive strength tests conducted in accordance with BS 1881 indicated that the fibres, when used at the recommended dosage rate of $0.90\,kg/m^3$, slightly increase the early strength gain of concrete. The fibres have no significant effect on the 28 day compressive strength of concrete cubes nor do they have any substantial effect on the flexural strength of concrete.

1.5.2.10 Shatter resistance.   Typically when concrete test cylinders fail in compression at ultimate load, there is an initial crack. Continued loading with plain concrete specimens causes the cylinders to fragment and fall apart. With polypropylene fibres present the concrete specimen holds together after maximum load without falling apart or shattering. Tests[1.16] show the ability of polypropylene fibre dosed concrete to remain intact and not to shatter after more than 10% additional compression as compared to plain concrete which completely shattered shortly after the first crack developed. This characteristic of polypropylene fibre dosed concrete is important in areas where there are impact or seismic concerns.

1.5.2.11 Impact resistance.   The addition of polypropylene fibres increases energy absorption/impact of a concrete slab. The fibres bridge the cracks that develop and thereby inhibit further crack growth. Therefore, whereas the ultimate tensile strength of fibre-dosed concrete does not increase appreciably, the tensile strain at rupture does. Where steel reinforcement is used in concrete the addition of fibres enhances the bond between the concrete and the reinforcing bars by inhibiting cracking in the concrete under bearing stress.

1.5.2.12 Abrasion resistance.   The introduction of polypropylene fibres into concrete results in a greater surface abrasion resistance compared to that of conventional concrete. Tests[1.16] have shown that the presence of fibres in a concrete mix reduces the amount of bleeding and assists in holding aggregate near the surface of fresh concrete so resulting in better surface integrity

1.5.2.13 Permeability.   Permeability is defined as the ease with which a fluid can flow through a solid. The addition of polypropylene fibres to a concrete slab reduces its water permeability owing to fibres interfering with the normal bleed channels and capillaries that are initially formed in the plastic state. With the reduction of cracking of the concrete resulting from the inclusion of fibres the penetration of water has been laboratory tested[1.16] to reduce by at least 50%. Figure 23 shows how increasing polypropylene fibre dosage decreases the permeability of the concrete.

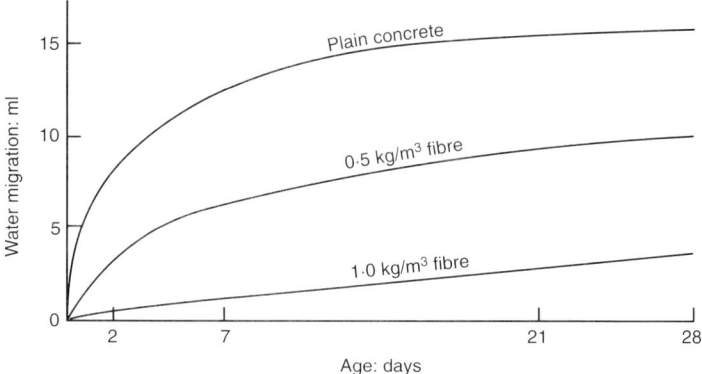

Fig. 23. Effect of polypropylene fibres on concrete permeability

1.5.2.14 Resistance to freeze/thaw. Fibre-dosed concrete has a significantly enhanced resistance to frost attack. There is some evidence to suggest that polypropylene fibres may be considered as an alternative to air-entrainment to obtain freeze/thaw resistance.

1.5.2.15 Chemical resistance. The presence of fibres does not alter the chemical resistance of concrete. Polypropylene is an inert and alkali resistant material and will not degrade in concrete.

1.5.2.16 Reducing corrosion of steel reinforcement. The addition of polypropylene fibres into a concrete slab significantly increases the protection of the steel reinforcement, within the slab, against corrosion. The reduction in permeability of the concrete is an attribute of prime importance with regard to the protection against corrosion. The high toughness index, which is the ability to sustain a load after initial crack, is also important in reducing the corrosion of reinforcing steel. This is due to the reduction of spalling of the concrete and the continued bond to the steel reinforcement.

1.5.2.17 Fire resistance. The inclusion of polypropylene fibres in a concrete mix does not affect the fire resistance of the finished concrete industrial hardstanding.

1.6 Reinforcing bars

Steel reinforcing bars should comply with BS 4461: 1978 (1984).[1.18] Information has been taken from BS 4461 to produce this Section. Steel reinforcing bars are delivered to site either in stock lengths, scheduled lengths or cut and bent to the specified shape. Stock lengths are usually

Table 1.17.   Preferred sizes of rebar[1.18]

Nominal diameter: mm	8	10	12	16	20	25	32	40

6 m or 12 m. Bars delivered to site should be tagged with an identification number or code relating to the hardstanding's reinforcing schedule or design drawings. The preferred diameters are shown in Table 1.17.

High yield grade 460 steel is use for industrial hardstanding concrete (characteristic tensile strength is 460 N/mm^2). Bars should be free from defects and, to ensure correct bonding to the concrete, there should be no loose rust, scale, grease or dirt present when the concrete is cast. Also if wire-guided vehicles are to be used, bars should be fixed at a sufficient depth to avoid interference with control signals. Bars are used where special reinforcement is needed, for example around gulleys or where there is a sudden change in level.

1.7   Surface finishes

1.7.1   Introduction

When selecting a specific type and method of finish a number of factors need to be considered. These factors include the type of traffic and loading the hardstanding will encounter, and the need for skidding, impact and chemical resistance. Usually, a brushed finish is sufficient to provide skidding resistance and durability. Figures 24 to 27 show some examples.

1.7.2   Curing compounds

Curing is a vital operation in the production of a durable concrete hardstanding surface. The main objective is to prevent early drying out of the surface, and therefore to allow full hydration to take place, resulting in a greater final strength and abrasion resistant concrete. Taking good care in curing reduces the risk of plastic cracking, dusting and drying shrinkage. Although traditional curing methods such as the application of wet hessian sacks and polythene sheets are still sometimes used, it is now common to specify a curing agent, often in the form of an acrylic polymer solution which impregnates the concrete surface forming a membrane. Application of the curing membrane is usually in the form of a spray onto the newly laid concrete after the moisture sheen has evaporated, or after final powerfloating. Many curing compounds combine the two functions of curing and sealing, thereby resulting in increased concrete surface protection.

Tests undertaken by a leading UK authority on the abrasion resistance of concrete hardstandings in relation to curing concluded that abrasion resistance can be greatly enhanced by the application of a resin-based sprayed-on membrane.[1.2] Polyethylene sheet proved to be the next most

Fig. 24.   Brushed finish provides sufficient skid resistance yet provides an acceptable ride

Fig. 25.   A brushed finish was applied but heavy rain occurred during the setting process. The resulting finish will be less durable and may be too slippery. Whether to leave such bays in place is often a significant point for discussion. This bay will need to be replaced or re-textured. Contractors should provide protective equipment such as tents

effective, followed by the wet hessian sack method and finally air alone. It was reported that the need for a good curing technique was particularly important when a high water/cement ratio was used.

1.8   Roller Compacted Concrete (RCC)
RCC is zero-slump concrete which is commonly used for industrial hardstandings in North America but which has yet to become widespread

Fig. 26.   This heavily trafficked 15 year old surface has lost its original laitance skin to reveal the aggregate. The skidding resistance may be lower, but otherwise this is acceptable and is normal in busy hardstandings

Fig. 27.   This surface was created by using wire tines rather than the more common wire brush. It provides additional skidding resistance but may be noisier

in the UK. It is particularly well suited to hardstandings subjected to heavy low-speed vehicular loading where surface appearance is of secondary importance and where standing water and a degree of undulation is acceptable. It is installed by a paving machine and is compacted by a vibrating roller. Joints are not provided within the hardstanding and cracks form in a random pattern which are sealed to prevent the ingress of moisture.

The characteristic compressive and flexural strengths normally specified are $25\,\mathrm{N/mm^2}$ and $3\,\mathrm{N/mm^2}$ respectively. It is impractical to entrain air into RCC so there remains the risk of damage through frost attack, although experience in North America indicates that the entrapped air in the material goes some way to providing frost protection.

There is more latitude in the selection of aggregates for RCC than for conventional pavement quality concrete since the dry mix prevents the segregation which may occur in conventional concrete. Well graded or dense graded aggregates, similar to those used in asphaltic concrete mixtures, have been found to be suitable. Aggregates with high fines contents (i.e. additional material finer than 75 microns) have been used successfully since the additional fine material produces a closed surface and reduces the amount of cement required.

2. Construction

2.1 Introduction

A correctly designed and constructed in-situ concrete hardstanding combines the advantages of hard wear, long life and the ability to carry heavy loads safely at low costs. The purpose of a hardstanding will vary according to application and each hardstanding requires its own individual characteristics including strength, skidding resistance, drainage falls and sometimes aesthetics. An important factor commonly taken into consideration is speed of construction and the savings which might accrue from fast-track construction. A fast-track method is decribed and compared with traditional installation methods.

2.2 Principal hardstanding issues

Four important issues to be considered in industrial concrete hardstandings are:

- durability
- joint configuration to avoid cracking
- drainage
- skidding resistance.

A unique feature in hardstandings is the relationship between design, construction and performance. All of the above issues are influenced by both design and construction.

2.3 Traditional construction methods[2.1]

Traditional long strip in-situ concrete hardstanding construction has been established as the conventional way of construction since the 1960s. The hardstanding is laid in a series of strips up to 6 m wide using timber or steel formwork. Often, every second strip is concreted initially, leaving infill unconcreted strips. The infill strips are concreted several days later using the originally laid strips as the formwork. A variation is to concrete a central strip initially, then install additional strips working away from the original strip so that all subsequent strips need formwork on one side only. Strips can be up to 60 m or more in length. The concrete may be placed and compacted in two or more layers, the upper layer being placed whilst the lower layers are still plastic. Compaction takes place using a combination of internal poker vibrators and twin-beam vibrating compactors running on the formwork or on the previously cast slabs. Some illustrations of construction are provided in Figs 28 to 31.

2.4   Large bay construction[2.1, 2.2]

Frequently, concrete hardstandings are installed using large bay or large pour methods so as to increase the daily output and to speed up the project. High workability concrete is used with a slump in excess of 150 mm. This is made possible by the addition of super-plasticizers. The high slump value allows concrete to be placed directly from a truck mixer and to be spread manually. The concrete almost self-levels such that a satisfactory level can be achieved by undertaking final adjustments based upon levels provided by laser transmitters. Any discrepancies from true level can be corrected using timber screed boards to bring the surface to the correct level. Compaction is achieved by lightweight screed beams and/or vibrating pokers. The principal disadvantages of the method are in the segregation of the aggregate with the larger particles sometimes sinking to the bottom of the slab and in the poor surface regularity often achieved. Segregation can lead to a high concentration of fines at the surface and consequent reduction in durability. This technique allows the use of steel mesh reinforcement which is usually laid out a day ahead of the placing of the concrete.

2.5   Laser screed construction[2.2, 2.3]

Laser-guided screeding machines were introduced to Europe and Scandinavia in the 1980s. The machines allow the cost effective installation of large concrete pours — outputs of over 2000 m^2 per day have been reported. Their principal innovation is that the level of the finished concrete is controlled automatically so the operator can focus upon achieving quantity. Laser-guided screeding machines combine state-of-the-art laser control systems with conventional mechanical screed mechanisms. The machines have four wheel drive, four wheel steer including 'crab' steer for awkward areas and are operated by one person seated at a point of maximum visibility (see Fig. 32).

Mounted on the twin axles, a circular fully slewing turntable carries a counter-balanced telescopic boom, typically having a 6 m reach on the end of which is attached a 3 to 4 m wide screed carriage assembly which

*Fig. 28.   The cropped end of each dowel bar is sawn off cleanly to prevent the
bar from locking the concrete in place rather than allowing it to slide*

*Fig. 29.   Concrete being discharged from the ready mix truck. The technician is
about to collect a sample for site testing. To avoid segregation, the concrete
should not form a deep pile but should be placed through the bay*

Fig. 30.   The concrete is being compacted by a poker vibrator. The operator is standing on the mesh reinforcement and is allowing the poker to reach to the bottom of the slab. Overvibrating leads to segregation and loss of both strength and durability

Fig. 31.   The bucket of a small excavator is used to place the concrete once it has been discharged from the ready mix truck. As well as avoiding the use of labour, it is less likely to disturb the reinforcement or the slip membrane

Fig. 32.   Laser-guided screeding machine ready to install a hardstanding. The receptors are at the top of the posts at each end of the carriage. Concrete is being delivered from the rear and is about to be pumped to the front of the machine carriage. The carriage compacts, levels and finishes the concrete which then requires a brushed or similar finish

comprises a plough, an auger to spread the concrete accurately and a vibrating beam for compaction. Test results have shown that such machines can compact concrete to depths in excess of 300 mm.

A self levelling laser transmitter is fixed at a visible point close to the work so as to project a 360° rotating beam across the working area. Depending on the type of transmitter, various inclinations of slabs can be achieved including level, single and dual falls. The level of the laser screeder is controlled by the laser beam which activates receivers mounted on the screed carriage assembly. During concreting the signals are relayed continuously to an on-board control box which automatically controls the level of the working screed head by adjusting the machine's hydraulic system. The laser transmitter rotates at 300 rpm so that the height of the screed carriage is adjusted five times per second.

2.5.1   Laser screed operation

Laser-guided screeding machines are used in conjunction with mixer trucks which place concrete 25–35 mm above the finished slab level. The positioning of the laser screeding machine, mixer trucks, slip membrane, joints and reinforcement requires careful organisation so as to attain maximum output. Once the concrete has been deposited from the mixer, the boom of the screeding machine is extended over the fresh pour and the screed carriage is lowered until the receivers lock onto the signal generated by the laser transmitter. The boom is then retracted, drawing the screed carriage towards the operator across the concrete so placing, compacting and screeding the slab simultaneously.

In one pass a laser-guided screeding machine can place compact and screed $20\,m^3$ concrete in under two minutes. Because of the geometry of the horizontal auger, screeding takes place from left to right with an overlap between sequential screeding runs of 300 mm so as to ensure optimum level and surface regularity across the entire slab. The only formwork required is that defining a day's concreting which in many cases will be the entire project. Often even that will not be required when, for example, the hardstanding is conctructed adjacent to buildings or other hardstanding concrete.

Output depends on site-specific details, type of reinforcement used (fabric can pose problems as the laser screeder tends to lift the mesh out of the concrete) and the speed at which the concrete supplier can deliver the concrete to site. Outputs of between $2000\,m^2$/day and $3000\,m^2$/day are normally possible. However, it should be recognized that placing $3000\,m^2$ of 200 mm thick concrete requires $600\,m^3$ of concrete which might require 100 concrete deliveries in one day.

2.5.2 Laser screeding level control

The screeding level of the machine's screed carriage is maintained by an automatic laser-controlled system. Laser receivers mounted at each end of the screed carriage detect the reference datum plane emitted by a laser level transmitter situated near the work area. The on-board control box checks and adjusts the screed carriage level in relation on the laser plane five times per second.

2.5.3 Advantages associated with laser screeding machines

Contractors using laser screeding machines have reported the following benefits as compared with traditional long strip construction.

- Higher strength, denser and more durable slabs.
- More accurate levels.
- Higher productivity.
- Ensures construction programmes are kept to time, enabling possibilities of earlier use of facilities.
- Damage-prone construction joints are kept to a minimum.
- Choice of mesh or steel fibre reinforcement.
- Larger areas of hardstanding can be placed, screeded, vibrated, compacted and left to cure in a single day. Outputs of 2000–$3000\,m^2$/day are achievable.

2.6   Finishing and curing processes

Once the hardstanding has been placed the final stages of finishing and curing are important to achieve an accurate and durable hardstanding with appropriate skidding properties.

2.6.1   Hardstanding finishing

Finishing is a critical and skilful operation which takes place after the hardstanding has been screeded. Finishing involves the use of a hand-held float which is drawn across a freshly laid slab to produce the correct level of surface laitance. The surface is then treated with a stiff brush to produce skid resistance. Once finishing has taken place the concrete is cured by the application of a resin-based reflective spray. Finishing processes are illustrated in Figs 33 to 36.

2.6.2   Curing

Curing is an essential part of the construction process. Some consider that the labour-intensive traditional method of covering the hardstanding with wet hessian for at least seven days is the most effective curing method. If water escapes too early the upper part of the slab will dry out quickly with slower rates of curing occurring below. This can lead to less durable concrete and slab curling. Curing time, and hence concrete strength, is affected by temperature, wind, rain and sun.

Fig. 33.   The surface is being trowelled with a steel float. By twisting the long handle, the operator can adjust the angle of the float in order to undertake the trowelling either by pushing or by pulling the float. This tool has been adapted by the fitting of a brush to be used to provide a brushed finish in a second operation

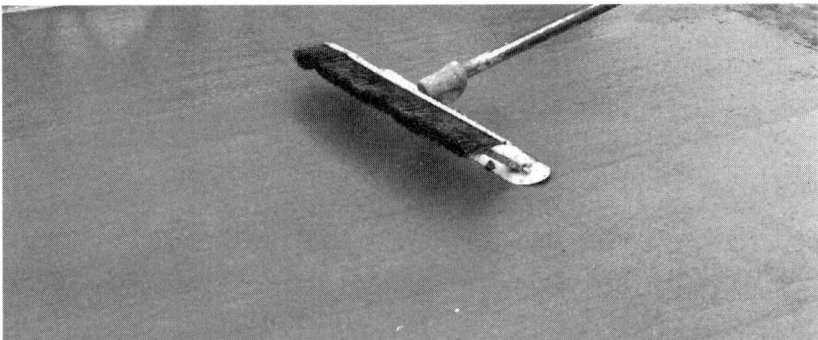

Fig. 34.   The operator is pulling the float towards himself and has angled it to avoid digging into the concrete. The brush is not being used at present but will be used to texture the surface once it has been floated

Fig. 35.   The edges of the bay are trowelled in those places where the main float failed to smooth the concrete. Care has to be taken to avoid overtrowelling which may well lead to an excess of surface laitance which might then spall off, leaving the concrete vulnerable

2.7   Slip membranes[2.1, 2.2]

2.7.1   The purpose of slip membranes

The purpose of slip membranes is to reduce the coefficient of friction between a concrete slab and its sub-base. It is not intended to act as a damp-proof membrane (DPM) though the material does allow the retention of some moisture. If a DPM were to be used then no moisture would escape downwards from the slab. Perforations cannot be entirely avoided so some moisture may escape into the sub-base. The slip layer reduces the coefficient of friction so allowing the concrete to move more easily as it shrinks during the curing process, hence reducing stresses in the slab, decreasing the possibility of cracking and assisting in the development of induced joints (see Fig. 37).

Fig. 36.   The combined tool is now being used to form the brush textured finish as the operator draws the brush towards himself. Care needs to be taken to maintain straight brush marks for the sake of appearance

Fig. 37.   Before concreting can commence, water is pumped away from the prepared bay. It can be difficult to remove all of the water and it is sometimes better to protect the bay from precipitation. This process is particularly important when a slip membrane has been installed

2.7.2   Slip membrane materials

At present there is no British Standard governing the use and type of slip membrane though polyethylene sheets are commonly used with thicknesses of 1000 gauge (250 microns) and 1200 gauge (300 microns).

2.7.3   Importance of sub-base with regard to slip membranes

Slip membranes are installed between the slab and sub-base immediately before the concrete is placed. They have to be strong enough to withstand construction traffic such as truck mixers. It is important to overlap the sheets by at least 200 mm and to use tape to ensure adjacent sheets do not move or separate. Care must be taken to avoid wrinkles and rips which may induce cracks in the strengthening slab.

If ruts are present in the sub-base surface and the slip membrane sheets are placed over them, voids may form beneath the slab when the concrete is placed. Weak spots will develop, leading to the possibility of cracking when load is applied. If the cross-section of the sub-base consists of rises and falls or peaks and troughs, the slip membrane might fail to do its task as the shape of the sub-base may cause interlock and therefore limit the horizontal movement of the slab as it shrinks. Stresses could increase and cracking ensue. It is important to ensure that the sub-base is as flat as possible and that care is taken in laying the slip membrane sheets.

2.7.4   Considerations in the design and use of slip membranes

There are no specific guidelines regarding the use of slip membranes and the designer has the choice of whether or not to use a slip membrane. The choice will be based on considerations of shrinkage, strength, curling and joint details. The purpose of the slip membrane is to allow the concrete slab to shrink freely and so reduce the levels of stress developed by restraint to movement. Allowing movement reduces the number of contraction joints required, but the joints which are provided may be more active. A consequence of eliminating the slip membrane is the development of interlock between the sub-base and the slab. Slab/sub-base friction is enhanced and the coefficient of friction increases from 0.2 to 0.7, so inhibiting the movement of the slab and requiring more contraction joints each moving by less.

Curling is caused by differential moisture loss between the surface and base of a concrete slab. When a slip membrane is used, the majority of moisture escapes from the slab through its upper surface which then shrinks and cures faster resulting in the slab's attempting to curl up at its edges. Eliminating a slip membrane allows moisture to escape through the base of the slab which leads to more even curing of the slab and so reduces curling stresses. If too much water escapes through the base of the slab, instead of curling upwards, the slab may develop hogging with the edges of a slab attempting to curl downwards. Blinding (i.e. providing fine material at the surface) the sub-base can reduce hogging by allowing moisture to escape from the slab whilst at the same time preventing it from flowing away through the sub-base. The ideal slip membrane would:

- allow the slab to move relative to the sub-base as it shrinks during curing enabling fewer joints to be constructed;
- be easy to handle and strong enough not to tear;
- be perforated so as to allow a controlled amount of moisture to escape and so eliminate curling.

2.8   Joint details

In the early life of a concrete hardstanding the slab will contract and be in danger of cracking. This can be controlled by an arrangements of joints. In hardstandings where operating conditions can permit the presence of cracks, for example in lightly loaded or permanantly covered slabs, the number of joints can be reduced and additional reinforcement provided. Particular attention should be paid to the alignment, setting and compaction of concrete at joints.

2.8.1   Movement joints

Movement joints are provided to ensure minimum restraint to movement caused by thermal and moisture changes in the slab. Movement joints are designed to allow the slab to contract. Expansion joints, or more correctly, full movement joints, are used in regions where temperature changes can be substantial in a short period of time and are reqired for slabs conctructed during cold weather in the UK.

2.8.1.1   Formed Doweled Construction (FDC) Joint.   This type of joint (Fig. 38) is provided at the ends and sides of a construction bay. It includes debonded dowel bars. To ensure the dowel bar and sleeve arrangement can move the bars are debonded using one of three techniques illustrated in Fig. 39. For slabs subjected to highway loading or less, 25 mm diameter dowels should be provided at 300 mm centres. For heavier loads, the dowels' cross-sectional area should be increased according to the loading. For very heavy loads, this can lead to an impractical dowel bar arrangement. In such cases, the neighbouring slabs should be designed according to the corner patch load method in Chapter 4 with no allowance being made for load transfer. Even then, the dowel bar details as set out above should be provided. Figures 40 and 41 illustrate some uses of dowel bars.

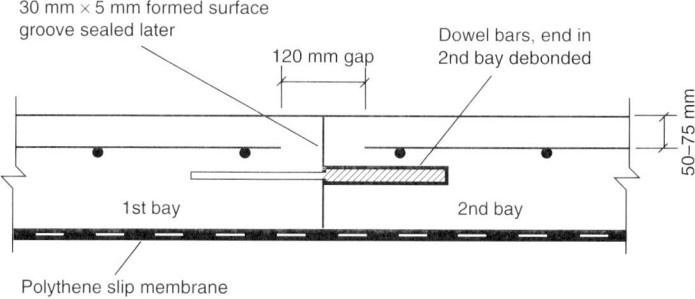

Fig. 38.   A Formed Doweled Contraction (FDC) joint

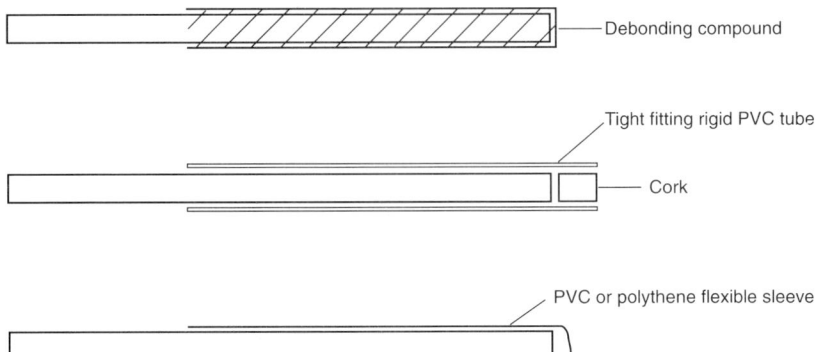

Fig. 39.   Three methods of debonding doweled joints

Fig. 40.   Dowel bars have been inserted into the timber formwork. Note the hole between the two dowels for cases where additional bars are required. The holes need to be blocked prior to concreting

2.8.1.2   Induced Doweled Contraction (IDC) Joint.   This type of joint is designed to allow movements as the concrete shrinks. It is used in construction bays which are designated large (any area greater than 1000 m^2) and which have been placed in a continuous operation, e.g. by a laser-guided screeding machine. This type of joint (Figs 42 and 43) is formed by crack induction (Section 2.8.3). The dowel bars enable vertical load transfer and are fixed prior to the placing of the concrete. The dowel bars should be 25 mm diameter at 300 mm centres. In the case of highway

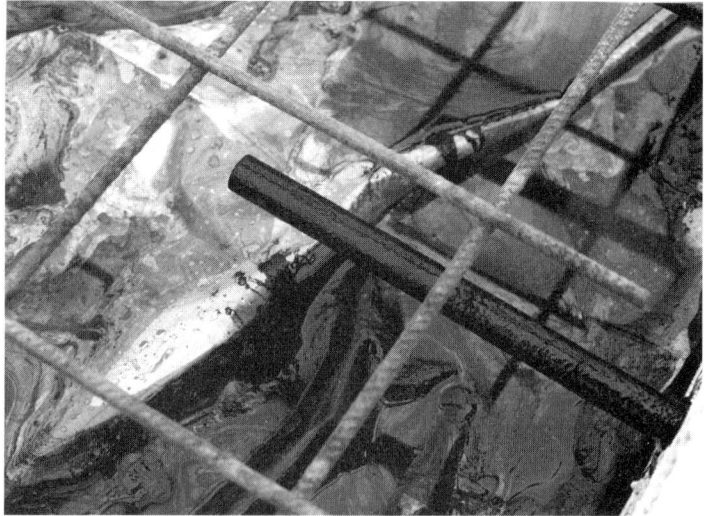

Fig. 41.   This half of the dowel bar has been coated with debonding compound to allow the neighbouring slabs to slide horizontally relative to each other

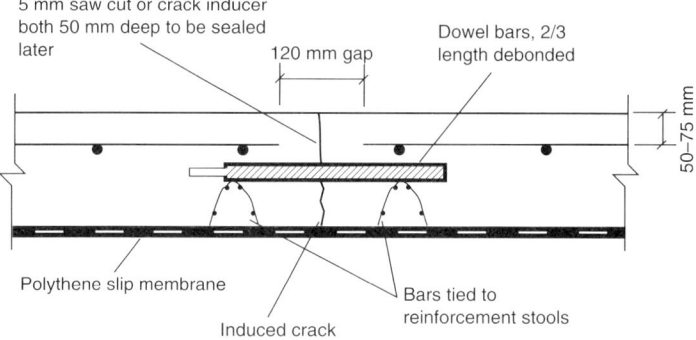

Fig. 42.   Induced Doweled Contraction (IDC) joint

vehicle loading or less, this is sufficient to ensure load transfer and design can proceed on the internal patch load condition. When loads exceed highway values, e.g. container handling plant, such an arrangement of dowels is inadequate and the slabs should be designed according to the corner patch method with no allowance for load transfer — the dowels should nonetheless be provided as set out above.

2.8.1.3   Induced Contraction (IC) Joint.   Fig. 44 shows the type of Induced Contraction joint used in large construction bays. It reduces the cost of joints by eliminating dowel bars and reduces the risk of joint

Fig. 43. In this induced joint, the concrete will be held together by a layer of A393 mesh at mid-depth so as to allow relative rotation whilst preventing horizontal relative movement

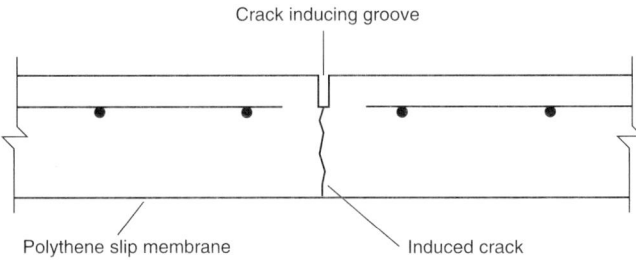

Fig. 44.   Induced Contraction (IC) joint

breakdown owing to poor workmanship in the placing of the dowel bars before concreting. Load transfer depends on aggregate interlock across the induction groove. Hardstandings incorporating this type of joint are suitable for lightly loaded applications where reinforcement requirements are minimal. They can be used for all situations up to and including highway loading.

2.8.1.4 Isolation joints.   Isolation joints or full movement joints, see Fig. 45, are used to permit the relative movement of a slab and neighbouring fixed structures, drains and perimeter fencing. They keep the concrete slab separate from fixed elements.

2.8.2 Tied joints
Tied joints hold two construction bays together using tie bars and are primarily used to restrain movement and contraction in the horizontal

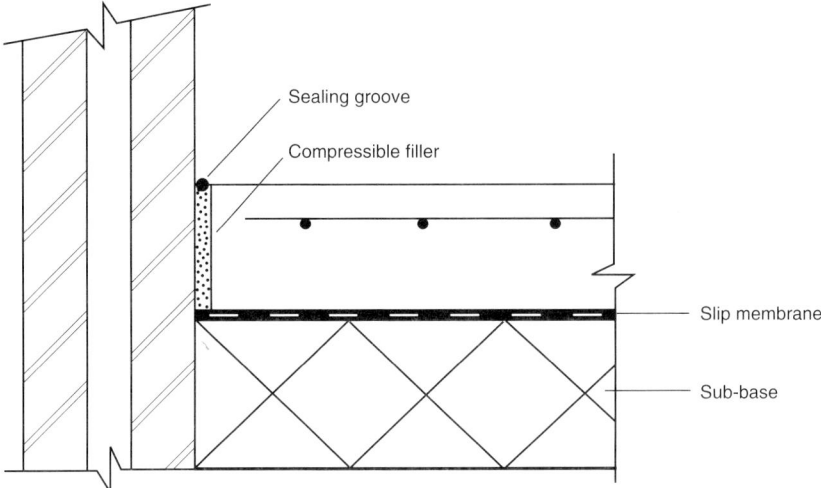

Fig. 45.   Isolation joint at a wall

Fig. 46.   Tie bars in a longitudinal joint. The bars provide very little bending resistance but prevent the slabs from moving horizontally whilst providing load transfer

plane (Fig. 46). The tie bars also act as a form of stress relief as they are bonded within the concrete and provide a large amount of strength across the joints. Tie bars are inserted in pre-drilled holes in the formwork as the concrete is placed. Where loads exceed highway loading, the arrangements of tie bars will be insufficient to transfer loads between slabs. In such cases, the slabs should be designed on the basis that there is no load transfer between neighbouring slabs, even though the tie bars are provided.

2.8.2.1   Formed Tied (FT) joint.   This type of tied joint (Fig. 47) is provided around the edges of a construction bay or at a stop-end

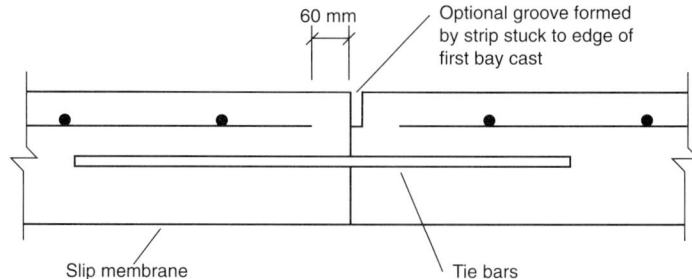

Fig. 47.   Formed Tied (FT) joint

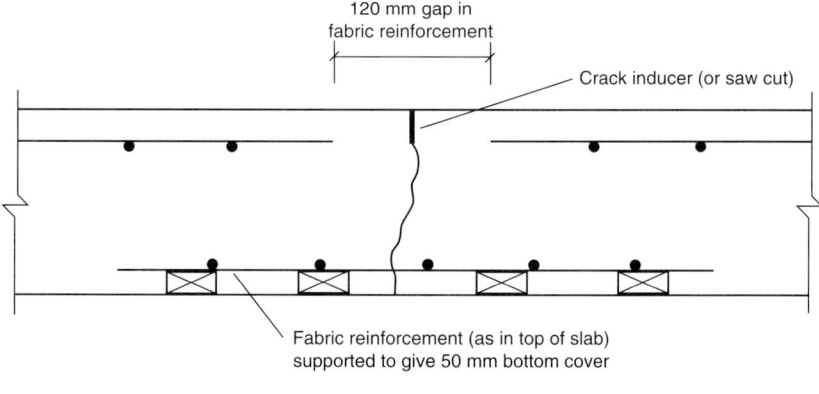

Fig. 48.   Induced Tied (IT) joint

commonly when constructing slabs using the long strip method and has proved successful in controlling movement at joints. A groove is provided to allow the joint to be sealed and is formed by a strip placed on the edge of the first bay cast.

2.8.2.2   Induced Tied (IT) joint.   Long strip slabs are prone to cracking. Using induced tied joints (Fig. 48), construction bays are reduced in size to reduce the possibility of cracking. As the hardstanding attempts to shrink it is restrained by the ties. Cracking may occur and a crack inducer is placed in the surface of the slab, encouraging the slab to crack at the joint.

2.8.3   Crack induction methods

2.8.3.1   Sawn joints.   The saw cut acts as a line of weakness which is incorporated into the slab at the position of the joint such that the slab will crack at that point owing to an increase in the tensile stress in the

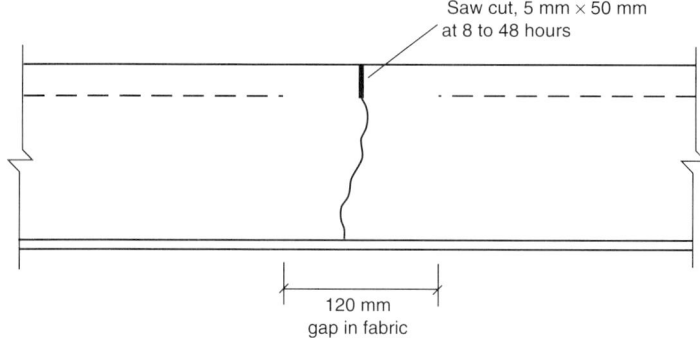

Saw cut, 5 mm × 50 mm
at 8 to 48 hours

120 mm
gap in fabric

Note: 50 mm suitable for slabs up to 200 mm thick.
For greater depths feature must be at least 1/4 depth

Fig. 49.   Crack induction joints using the saw cut technique

Fig. 50.   A slot or groove has been sawn to a depth of 50 mm across the bay. As yet, the concrete has not cracked. Cracking takes place over a period of several weeks

remaining depth of slab. Saw cuts are formed when the concrete has gained sufficient strength to withstand the effects of a concrete saw but not so much that the effect of sawing would damage the hardstanding. Sawn joints are particularly durable. They are expensive and can cost as much as ten times as much as wet formed joints. The saw cut breaks the upper layer of reinforcement and the groove has a depth of 40 mm or 50 mm (Fig. 49). As well as the joint being the most durable crack induction form, a sawn joint is also very serviceable with no difference in level at each side of the cut. This aids flatness which is important on heavily trafficked hardstandings. Figures 50 to 52 show some examples.

2.8.3.2   The timing of forming sawn joints.   Joints should be cut when the concrete has gained sufficient strength to support the weight of the

Fig. 51.   Care has to be taken both in the timing of the sawn joints and in avoiding spalling of the new concrete, particularly at the end of the saw cut

Fig. 52.   This crack formed at an induced joint a few days after the concrete was cast. In this case, the slot or groove was formed in the wet concrete

cutting equipment but before it has gained sufficient strength that sawing might loosen or pull out aggregates or steel fibre reinforcement. A suggested sawing time scale is between 24 and 48 h after initial concrete set. This leaves a time window, as shown Fig. 53 assuming the following.

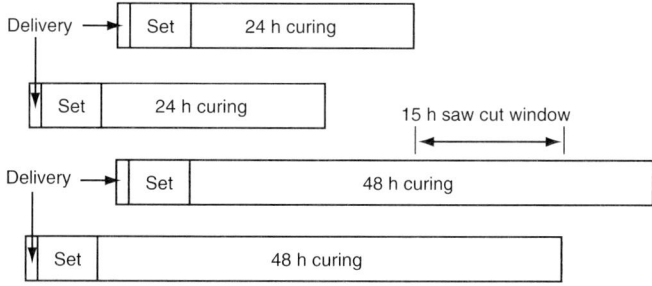

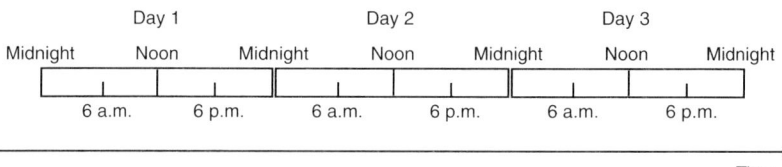

Fig. 53.   *The timing of forming sawn joints*

- Concrete is mixed 1 h before placing.
- First concrete is mixed at 7.00 a.m.
- Last concrete is mixed at 4.00 p.m.
- Concrete is placed between 8.00 a.m. and 5.00 p.m. on day 1.
- Concrete takes 6 h to reach initial set.
- Therefore, earliest initial set = 1.00 p.m.; latest initial set = 10.00 p.m.

If the concrete is to be sawn between 24 and 48 h after the initial set, the first cut can be performed 24 h after the latest initial set so as to ensure that all of the concrete has gained sufficient strength. All saw cutting must be finished 48 h after the earliest initial set. In this example, saw cutting can commence at 10.00 p.m. on day 2 and must be finished by 1.00 p.m. on day 3. This gives 15 h sawing time.

2.8.3.3 Other methods of crack induction.   Figure 54 shows traditional crack inducers involving the use of metal or plastic strips which are inserted into the wet concrete after placing. The upper section of the strip is removed once the concrete has hardened, leaving the lower section in the slab to form the crack inducer. The insert (Fig. 54(*b*)) is shaped so as not to disrupt the concrete surface when inserted. It is strong enough to be able to be pushed into the concrete vertically. The Zip-Strip is used more commonly in the US and comprises a plastic extrusion made from two identical parts to form a T-shape. The rigidity of the plastic section allows

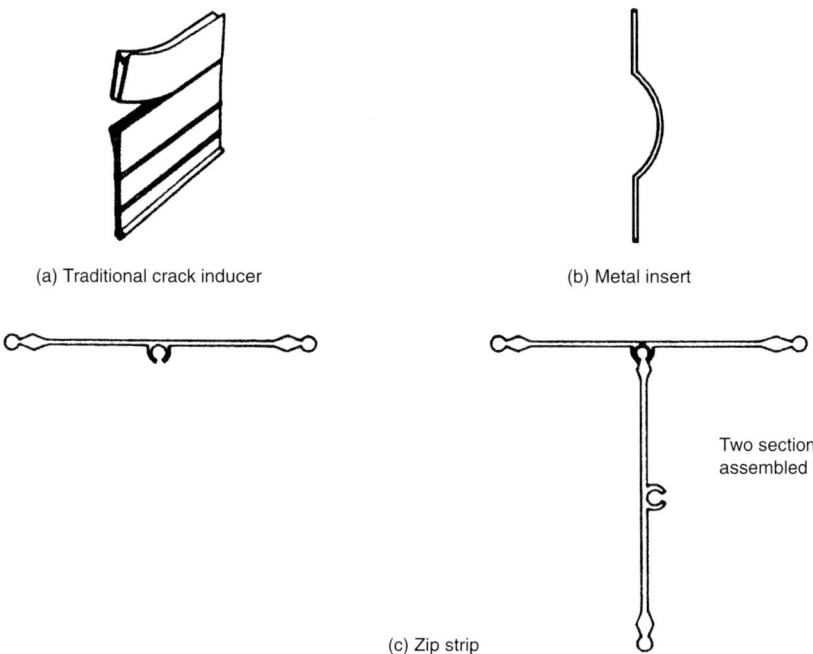

(a) Traditional crack inducer

(b) Metal insert

Two section:
assembled

(c) Zip strip

Fig. 54.   Traditional crack inducers

the vertical section to be pushed into the concrete and the top section is removed for reuse. Immediately following concreting, a groove is formed using a bricklayers' trowel against a string line to ensure accuracy. Once the inducing strip is in place further compaction and vibration is necessary to ensure that no air has been introduced into the concrete.

2.8.4   Position of joints[2.4]

In the case of mesh reinforcement, joint spacing in each of the two orthogonal directions is proportional to the weight of the bars running at right angles to the joints being designed. Figure 55[2.4] shows the relationship between mesh weight and joint spacing. Note that it is frequently the case that joints will be spaced at different centres in each orthogonal direction. Transverse joints may be widely spaced with longitudinal joints being kept at say 4 m to 5 m spacings to facilitate construction. For such hardstandings, long mesh reinforcement may be specified with the main bars running parallel to the longitudinal joints.

Table 2.1 shows joint spacings for various types of steel fibre reinforced concrete. The joint spacings in the table have been used successfully for many years in the UK. Some consider that spacings can be greater than 12 m for steel fibre reinforced hardstandings. Whilst 14 m

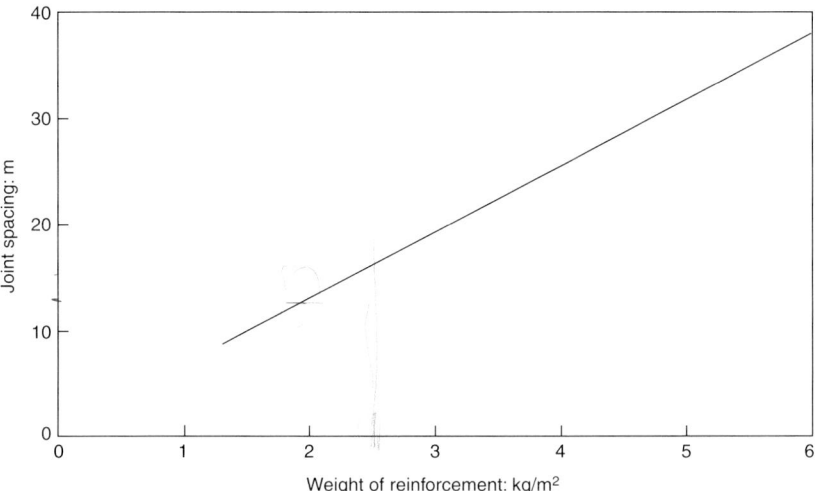

Fig. 55.   Joint spacing in relation to weight of steel mesh[2.4]

Table 2.1.   Common concretes and suggested joint spacings

Concrete type	Joint spacing: m
Plain C30 concrete	6
20 kg/m^3 ZC 60/1.00 steel fibre reinforcement C30 concrete	7
30 kg/m^3 ZC 60/1.00 steel fibre reinforcement C30 concrete	8
40 kg/m^3 ZC 60/1.00 steel fibre reinforcement C30 concrete	10
Plain C40 concrete	6
20 kg/m^3 ZC 60/1.00 steel fibre reinforcement C40 concrete	8
20 kg/m^3 ZC 60/1.00 steel fibre reinforcement C40 concrete	10
20 kg/m^3 ZC 60/1.00 steel fibre reinforcement C40 concrete	12

joint spacings will probably be acceptable for some applications, the additional movement which would occur at joints might lead to loss of aggregate interlock and joint degradation. In any project it is necessary to develop a joint layout identifying all practical issues prior to the placing of concrete. Figure 56 shows the possible consequences of incorrect joint spacings.

Fig. 56.   This mid-bay transverse crack has formed as a result of the joints being too widely spaced

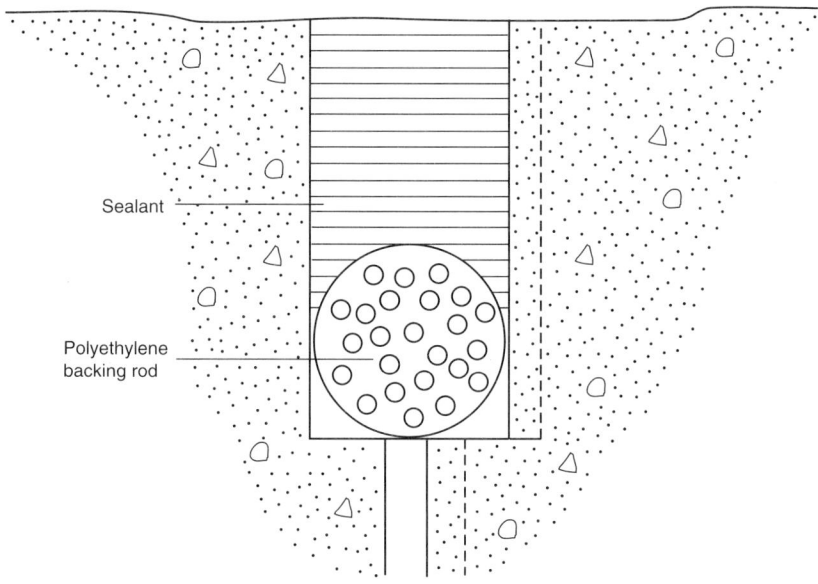

Sealant

Polyethylene
backing rod

Fig. 57.   Details of joint sealing groove

2.8.5   Joint sealing

The purpose of sealing joints is to prevent detritus, water and other debris entering the joint. The sealant must be able to withstand the strain of the opening of the joint resulting from contraction of the slab and remain fixed to the faces of the groove sides (Fig. 57). For hardstandings which

Fig. 58.   Sealed joint after several years' heavy traffic

are to be trafficked by rigid-tyred vehicles and where joint widths are greater than 5 mm, it may be necessary to use a strong semi-rigid sealant such as pouring grade epoxy or grout in order to provide support to the edges of the joint. The low elasticity of such sealants might cause them to fail when applied to an active joint. For this reason, they are applied at a later date when shrinkage movements have occurred. Alternatively, more flexible mastic-based materials can be used. In slabs where joint spacings exceed 10 m, it is particularly important to carry out the joint sealing. With deep crack inducing grooves a polyethylene backing rod may be placed in the lower part of the sealing groove to reduce the amount of sealant required (Fig. 57).

Durability can be improved with the use of polysulphide sealants. When appearance is of concern, gun-grade materials can be used, especially for narrow grooves. Sealants should be inspected and maintained regularly. Examples of joint sealing are shown in Figs 58 and 59.

2.9   Drainage

Hardstanding drainage needs to be considered alongside the design of the area since bay sizes, slopes (falls) and joint details may all influence the drainage system. The first issue to consider is where the precipitation

Fig. 59.   Spalling has occurred at this sealed joint. The arris is a vulnerable position where an absence of large aggregate in the concrete can lead to this type of deterioration

falling on the hardstanding will go when it leaves the site. The designer will need to liase with the local drainage authority to establish whether the downstream drainage system has sufficient existing capacity to accept the drainage from the proposed site. Also, the authority may require that a petrol interceptor be installed to remove any fuel or other light contaminents which become mixed with the surface drainage. In cases where there is insufficient downstream capacity, it may be necessary to introduce a detention system which will hold storm water until sufficient capacity becomes available. Large diameter pipes often constitute a cost effective detention system. In this case, the diameter is designed to achieve the storage volume required rather than the flow rate.

Surface drainage can be by gulleys or by a linear drainage system, which may include 'built-in' falls. Calculations are required for the number of discharge points, but a commonly adopted rule of thumb is that a gulley is required for each 300 m² of hardstanding with 150 mm diameter pipes leading to progressively higher capacity pipes as the water progresses towards the existing drainage system.

Surface falls are often specified as a compromise between the ideal 1 in 40 (2.5%) which ensures rapid removal of surface water and the flat horizontal surface which often suits operations best. Falls of 1 in 80 to 1 in 100 are common. Figure 60 shows a typical drainage arrangement for a medium sized hardstanding. The gulleys have been located in the slab

Fig. 60.   Drainage gulley positioned within a slab. Note the channel formed in the concrete to encourage the water to reach the gulley grating

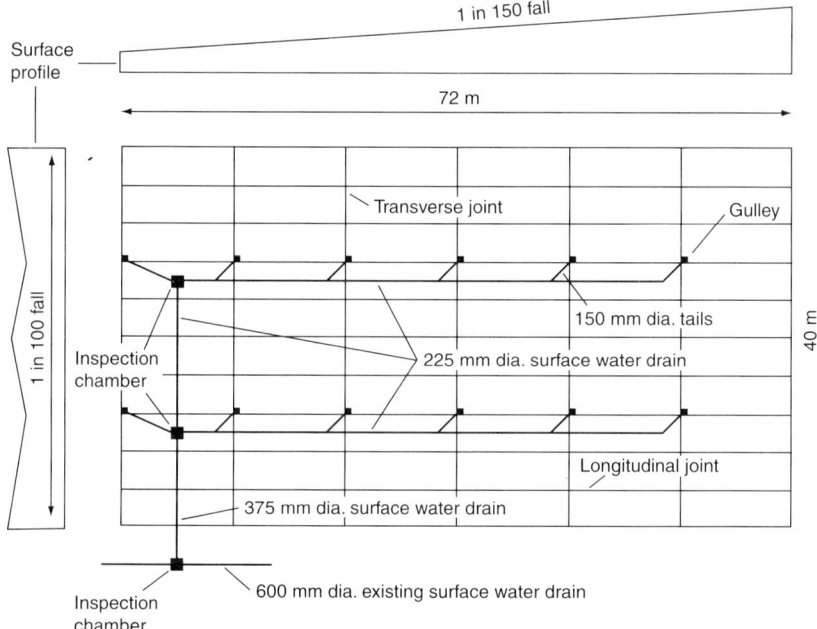

Fig. 61.   Typical hardstanding showing relationship between drainage and joints. This is a long strip hardstanding

corners. Care needs to be taken in the detailing of the adjacent concrete and some designers prefer to locate gulleys near the centre of slabs. Wherever a gulley is located there will be a reduction in slab strength. They should be located away from the zones of repeated heavy loading or container corner castings.

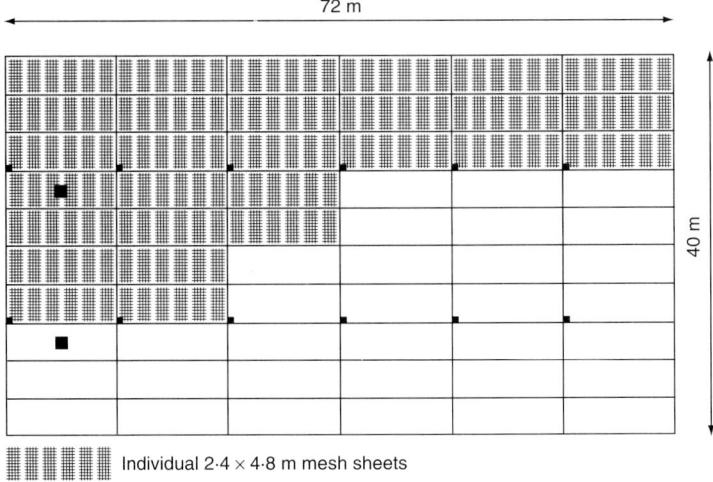

Fig. 62. *The positioning of steel mesh in a typical long strip industrial hardstanding. Each bay has six sheets of mesh with 200 mm overlap*

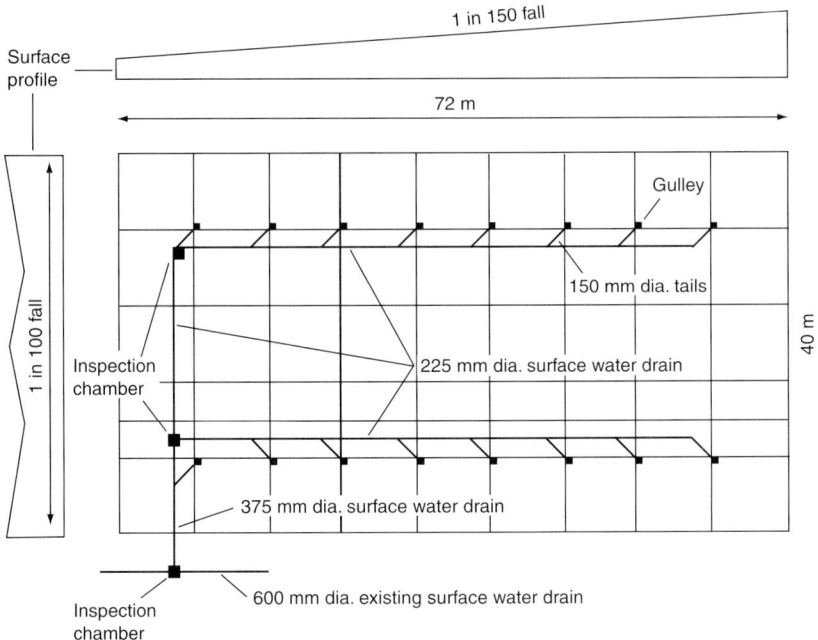

Fig. 63. *Typical hardstanding showing relationship between drainage and joints. This is an example of a single pour hardstanding in which the joints have been formed by sawing grooves*

2.10   Hardstanding construction case study

Figure 61 illustrates a typical long strip hardstanding showing the joints, the falls and the drainage. If the operational characteristics of the area are determined at design time, the gulleys can be located away from the predominant loading. The contractor might opt to construct alternate strips, each with formwork at both sides, then construct the intermediate strips, using the existing concrete as formwork.

Figure 62 shows the way in which the individual sheets of 2.4 m × 4.8 m mesh are located within each bay. In this case, each sheet will have to be shortened from 4.8 m to just less than 4 m to fit the 4 m wide strips. Figure 55 shows that if A393 mesh is used, the distance between free joints can be up to 40 m. This means that all of the longitudinal joints can be tied and only one full movement transverse joint is required — the remaining transverse joints can be tied with a mid-depth sheet of mesh, or they can be eliminated.

Figure 63 shows the same hardstanding constructed using a laser-guided screed machine. By including steel fibres in the concrete, the mesh can be omitted, so facilitating construction. It is usual to use the same joint spacing in both orthogonal directions. Increasing the weight of steel fibres increases the joint spacing and decreases the slab thickness required. The most cost effective solution is not obvious. The additional cost of fibres can be offset by a reduction in joint spacing and in a reduction in joint construction costs.

3. Loading

3.1 Wheel load value

The loading regime to be used with the equations in the case of patch loads or the design chart in the case of point loads in Chapter 4 is rationalized to a single equivalent load describing the actual regime. When the design process is started there is usually no unique load value that characterizes the operational situation. Consequently it is necessary to gather information known about the loading environment in order to derive the equivalent single load to be used with the design procedure. Firstly, information regarding the types of loads that can be expected is given with factors that should be considered. This is followed by a rational method of deriving the single equivalent pavement load required for use with the Design Chart through proximity and dynamic factors.

Many industrial hardstandings are loaded by highway vehicles or by less onerous plant and equipment. The maximum legal axle load on a UK highway is 11 500 kg, but surveys indicate that some vehicles are overloaded. It is recommended that in the absence of more accurate data, industrial hardstandings trafficked by highway vehicles or by lighter plant are assumed to be loaded by axles of weight 14 000 kg. This takes into account overloading and dynamics, but not wheel proximity. It may be that an industrial hardstanding can be designed for relatively few such vehicles, say 5% of the total vehicles expected to traverse the busiest point in the slab. Figures 64 to 66 show some examples of wheel loads common on industrial hardstandings.

Where loading exceeds highway levels, the usual reason is the handling of containers by off-road plant such as straddle carriers or front lift trucks (see Figs 67 and 68). In such cases, the value of the design wheel load depends upon the range of container weights being handled. Design should be based upon the critical load, which is defined as the load whose value and number of repetitions leads to the most pavement damage. Relatively few repetitions of a high load value may inflict less damage than a higher

Fig. 64.   Four-wheel straddle carriers apply wheel loads in excess of 20t

Fig. 65.   Front lift trucks handling heavy 40ft containers apply 100t or more through the front axle

Fig. 66.   Highway trailers may have three axles, each applying 11t. The small steel 'jockey' wheels may apply even greater load when the trailer is parked. A trailer is never moved using its jockey wheels and, in many cases, a plate is provided instead

Fig. 67.   Straddle carriers are preferred to front lift trucks when significant travel distances are involved and where two- or three-high stacking occurs

Fig. 68.   Because these containers are stacked four high, a front lift truck (FLT) is used. In this case, the containers are empty so a smaller truck is required

number of lesser load values. The entire load regime should be expressed as a number of passes of the critical load. Evaluation of the critical load and the effective number of repetitions of that load is as follows.

Table 3.1 shows the distribution of container weights normally encountered in UK ports for different proportions of 20 ft and 40 ft containers. Where local data are available, these can be used in place of Table 3.1. For each of the container weights shown in Table 3.1, the damaging effect caused when plant is handling containers of that weight can be calculated from the following equation:

$$D = (W/12000)^{3.75}(P/0.8)^{1.25}N \tag{3.1}$$

where D is the damaging effect, W is the wheel load corresponding to the specific container weight (in kg), P is the tyre pressure (in N/mm^2) and N is the percentage from Table 3.1.

Table 3.1. Percentages of containers of different weights for five different combinations of 40 ft to 20 ft containers derived from statistics provided by UK ports

Container weight: kg	Proportion of 40 ft to 20 ft containers				
	100/0	60/40	50/50	40/60	0/100
0	0.00	0.00	0.00	0.00	0.00
1000	0.00	0.00	0.00	0.00	0.00
2000	0.00	0.18	0.23	0.28	0.46
3000	0.00	0.60	0.74	0.89	1.49
4000	0.18	1.29	1.57	1.84	2.95
5000	0.53	1.90	2.25	2.59	3.96
6000	0.98	2.17	2.46	2.76	3.94
7000	1.37	2.41	2.67	2.93	3.97
8000	2.60	3.05	3.16	3.27	3.72
9000	2.82	3.05	3.11	3.17	3.41
10 000	3.30	3.44	3.48	3.52	3.66
11 000	4.43	4.28	4.24	4.20	4.04
12 000	5.73	5.24	5.12	4.99	4.50
13 000	5.12	4.83	4.76	4.69	4.41
14 000	5.85	5.38	5.26	5.14	4.67
15 000	4.78	5.12	5.21	5.29	5.63
16 000	5.22	5.58	5.67	5.76	6.13
17 000	5.45	5.75	5.83	5.91	6.21
18 000	5.55	5.91	6.00	6.10	6.46
19 000	6.08	6.68	6.83	6.98	7.58
20 000	7.67	8.28	8.43	8.58	9.19
21 000	10.40	8.93	8.56	8.18	6.72
22 000	9.95	7.60	7.02	6.43	4.08
23 000	5.53	4.31	4.00	3.69	2.47
24 000	2.75	1.75	1.50	1.25	0.24
25 000	0.95	0.63	0.55	0.47	0.15
26 000	0.67	0.40	0.33	0.27	0.00
27 000	0.72	0.43	0.36	0.29	0.00
28 000	0.53	0.32	0.27	0.21	0.00
29 000	0.43	0.26	0.22	0.17	0.00
30 000	0.28	0.17	0.14	0.11	0.00
31 000	0.03	0.02	0.02	0.01	0.00
32 000	0.03	0.02	0.02	0.01	0.00
33 000	0.00	0.00	0.00	0.00	0.00
34 000	0.05	0.03	0.02	0.02	0.00

The container weight leading to the greatest value of D is the critical weight container and all subsequent wheel load calculations should be based upon this load. Experience indicates that when the containers being handled comprise 100% 40 ft containers, the critical load is commonly 22 000 kg and when 20 ft containers are being handled, the critical load is

20 000 kg. In general, mixes of 40 ft/20 ft containers have a critical container weight of 21 000 kg. These values may be used in preliminary design studies. The number of repetitions to be used in the design can be calculated accurately using a load value weighted system. However, if the total number of repetitions calculated solely from operational data is used, a negligible error will be generated. In the case of pavements trafficked by highway vehicles, an equivalent load of 140 kN may be used.

3.2   Tyres

The contact area of a tyre of handling plant is assumed to be circular with a contact pressure equal to that of the tyre pressure. Some larger items of plant may be fitted with tyres for operating over soft ground. When such tyres travel over concrete the contact area is not circular and the contact stress under the tread bars is greater than the tyre pressure. This has little effect in the case of in-situ concrete. Some terminal trailers are fitted with solid rubber tyres. The contact stress depends upon the trailer load but a value of $1.7 \, N/mm^2$ is typical and the higher pressure is dispersed satisfactorily through the pavement.

3.3   Dynamics

The effects of dynamic loading induced by cornering, accelerating, braking and surface unevenness are taken into account by the factor f_d. Where a section of a pavement is subjected to dynamic effects the wheel loads are adjusted by the factors given in Table 3.2 as explained in the note to the table.

3.4   Lane channelization

Plant movements over a wide pavement do not follow exactly the same course, but wander to one side or the other. If there are lane markings with the lane approximately the same width as the plant, then channeling becomes significant. As the lane width increases relative to the width of the plant, channelization becomes less significant with the less channelized travel causing an ironing out effect more evenly over the area. For straddle carriers stacking containers in long rows the wheels are restricted to very narrow lanes and consequently severe rutting may take place (Fig. 69). In such a case the operation techniques of the plant in that area should be reviewed periodically.

3.5   Static loading

Static loads from corner casting feet apply very high stresses to the pavement (Fig. 70). These stresses can be taken by the concrete but some superficial damage may occur to the surface. In extreme cases, this

Table 3.2. Table of dynamic load factors (f_d). Static loads are increased by the percentage figures in the table

Condition	Plant type	f_d: %
Braking	Front lift truck	± 30
	Straddle carrier	± 50
	Side lift truck	± 20
	Tractor and trailer	± 10
Cornering	Front lift truck	40
	Straddle carrier	60
	Side lift truck	30
	Tractor and trailer	30
Acceleration	Front lift truck	10
	Straddle carrier	10
	Side lift truck	10
	Tractor and trailer	10
Uneven surface	Front lift truck	20
	Straddle carrier	20
	Side lift truck	20
	Tractor and trailer	20

Note. Where two or three of these conditions apply simultaneously, f_d should take into account multiple dynamic effects. For example, in the case of a front lift truck cornering and accelerating over uneven ground, the dynamic factor is 40% + 10% + 20% i.e. 70%, so that the static wheel load is increased by 70%. In the case of braking, the dynamic factor is additive for the front wheels and subtractive for rear wheels. In the case of plant with near centrally located wheels (e.g. straddle carriers), braking and accelerating dynamic factors to be applied to the near central wheels are reduced according to geometry

Fig. 69.   When operating within container stacks, a straddle carrier tracks the same length of the slab each pass

Fig. 70.   Failure of concrete slab in the vicinity of container corner castings. When the deformation exceeds 12 mm, the containers rest on their underside and the slab load becomes small. This is unacceptable from the structural capacity of the containers

damage, in conjunction with frost attack, might be sufficient to introduce structural implications.

3.5.1   Container corner casting load values

Containers are usually stacked in rows or blocks and, until recently, usually stacked no more than three high, with a maximum of five high. However, in recent times containers have been stacked up to eight high in a few locations and this may become more common in future. Corner castings measure 178 mm × 162 mm and frequently they project 12.5 mm below the underside of the container. Table 3.3 gives the maximum loads and stresses for most stacking arrangements. Since it is unlikely that all containers in a stack will be fully laden the maximum gross weights will be reduced by the amounts shown. The values shown in Table 3.3 can be used directly in the Design Chart. In the case of empty containers, pavement loads can be calculated on the basis that 40 ft containers weigh 3000 kg and 20 ft containers weigh 2000 kg.

3.5.2   Trailer dolly wheels

There are often two pairs of small or 'dolly' wheels on trailers which are 88 mm wide × 225 mm in diameter. When the trailer is parked, the contact

Table 3.3. Pavement loads from stacking full containers

Stacking height	Reduction in gross weight: %	Contact stress: (N/mm^2)	Load on pavement for each stacking arrangement: kN		
			Singly	Rows	Blocks
1	0	2.59	76.2	152.4	304.8
2	10	4.67	137.2	274.3	548.6
3	20	6.23	182.9	365.8	731.5
4	30	7.27	213.4	426.7	853.4
5	40	7.78	228.6	457.2	914.4
6	40	9.33	274.3	548.6	1097
7	40	10.9	320.0	640.0	1280
8	40	12.5	365.8	731.6	1463.2

Fig. 71.   These trailer dolly wheels have indented the bituminous material surfacing

area of each wheel is approximately 10 mm × 88 mm and stresses are 40 N/mm^2 (see Fig. 71). Some trailers have pivot plates which measure 150 mm × 225 mm and produce contact stresses of 2.0 N/mm^2; this is sufficiently low to cause no difficulties within the concrete.

3.6   Wheel proximity factors

The active design constraint is horizontal tensile stress at the underside of the slab in the case of internal and edge loading, and horizontal tensile stress at the surface of the slab in the case of corner loading. If one wheel only is considered, the maximum horizontal tensile stress occurs under the centre of the wheel and this is reduced with distance from the wheel. If two wheels are sufficiently close together, the stress under each wheel is

Table 3.4. Wheel proximity factors

Wheel spacing: (mm)	Proximity factor for effective depth to base of:		
	1000 mm	2000 mm	3000 mm
300	1.82	1.95	1.98
600	1.47	1.82	1.91
900	1.19	1.65	1.82
1200	1.02	1.47	1.71
1800	1.00	1.19	1.47
2400	1.00	1.02	1.27
3600	1.00	1.00	1.02
4800	1.00	1.00	1.00

increased by a certain amount owing to the other wheel. The method described here should be used when the California Bearing Ratio (CBR) of the subgrade is known. In cases where the modulus of subgrade reaction (K) is known more accurately than the CBR, the proximity factor calculation method presented in Chapter 4 should be used.

Wheel loads are modified by the appropriate proximity factor taken from Table 3.4. These factors are obtained as follows. If the wheel proximity were not considered, the relevant stresses would be the radial tensile stress directly beneath the loaded wheel. If there is a second wheel nearby, it generates tangential stress directly below the first wheel. This tangential stress is added to the radial stress contributed by the primary wheel. The proximity factor is the ratio of the sum of these stresses to the radial tensile stress resulting from the primary wheel. The following equations are used to calculate the stress:

$$\sigma_R = \frac{W}{2\pi}\left(\frac{3r^2 z}{\alpha^{5/2}} - \frac{1 - 2\nu}{\alpha + z\,a^{1/2}}\right) \tag{3.2}$$

$$\sigma_T = \frac{W}{2\pi}(1 - 2\nu)\left(\frac{z}{\alpha^{3/2}} - \frac{1}{\alpha + z\,a^{1/2}}\right) \tag{3.3}$$

where:

σ_R = radial stress
σ_T = tangential stress
W = load
r = horizontal distance between wheels
z = depth to position of stress calculations
ν = Poisson's ratio
a = $r^2 + z^2$.

When more than two wheels are in close proximity, the radial stress beneath the critical wheel may have to be increased to account for two or more tangential stress contributions. Table 3.4 shows that the proximity factor depends on the wheel spacing and the effective depth of the slab. The effective depth can be approximated from the following formula and represents the depth of the slab should the slab have been constructed from subgrade material.

$$\text{Effective depth} = 300\sqrt[3]{\frac{35\,000}{CBR \times 10}} \qquad (3.4)$$

where CBR is the California Bearing Ratio of the subgrade.

As an example, consider a front lift truck with three wheels at each end of the front axle. The critical location is beneath the centre wheel. Suppose a hardstanding were designed on ground with a CBR of 7% and the wheel lateral centres were 600 mm. From the formula, the approximate effective depth of the slab is:

$$\text{Effective depth} = 300\sqrt[3]{\frac{35\,000}{7 \times 10}} = 2381\,\text{mm} \qquad (3.5)$$

By linear interpolation from Table 3.4 the proximity factor is 1.86. This should be applied twice for the central wheel. This means that the effective single load is scaled up by 0.86 twice, i.e. $1 + 0.86 + 0.86 = 2.72$. Note that this is approximately 10% less than three so that this type of wheel arrangement effectively reduces slab load by 10%. For wheels bolted side by side where the wheel centres are separated by less than 300 mm, the entire load transmitted to the slab through one end of the axle can be considered to represent the wheel load. An investigation of the actual equivalent wheel load indicates that the actual equivalent wheel load is approximately 1.97 times one wheel load when there are two wheels bolted together at an axle end.

3.7   Wheel load calculations for handling plant

The following formulae are for guidance only and relate to plant having wheel configurations as illustrated in the diagrams. In cases where plant has an alternative wheel configuration, the loads can be derived from first principles, following a similar approach. In many cases wheel loads are provided by plant manufacturers and if this is the case, those values are to be preferred. For each pass of the plant, a specific spot in the slab is loaded by all of the wheels at one side of the plant. Therefore in the wheel load calculations, only one side of the plant is considered. In the case of asymmetrical plant, the heavier side should be chosen.

3.7.1   Front lift trucks and reach stackers (Figs 72–74)

In the examples illustrated in Figs 72–74

$$W_1 = f_d \left(\frac{A_1 W_c + B_1}{M} \right) \qquad (3.6)$$

$$W_2 = f_d \left(\frac{A_2 W_c + B_2}{2} \right) \qquad (3.7)$$

where:

$W_1 =$ load on the front wheel (kg)
$W_2 =$ load on the rear wheel (kg)

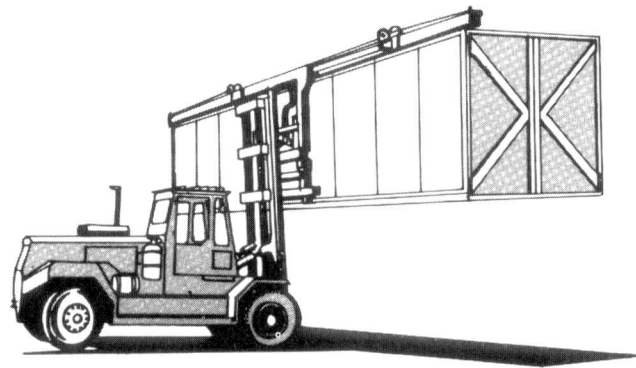

Fig. 72.   Front Lift truck handling 40 ft container

Fig. 73.   Reach stacker handling 40 ft container

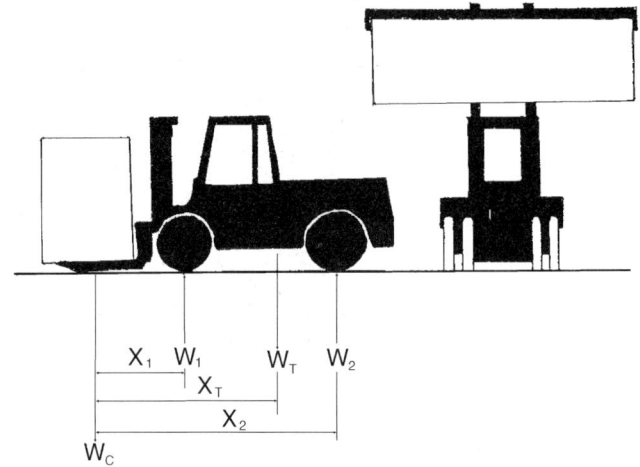

Fig. 74. Dimensions and weights used in wheel load calculations

W_c = weight of the container (kg)
M = number of wheels on the front axle (usually 2, 4 or 6)
f_d = dynamic factor.

A_1, A_2, B_1 and B_2 are given by:

$$A_1 = \frac{-X_2}{X_1 - X_2} \tag{3.8}$$

$$A_2 = \frac{X_1}{X_2 - X_1} \tag{3.9}$$

$$B_1 = \frac{W_T(X_T - X_2)}{X_1 - X_2} \tag{3.10}$$

$$B_2 = \frac{W_T(X_T - X_1)}{X_2 - X_1} \tag{3.11}$$

where X_1, X_2 and W_T are shown in Fig. 74 and W_T is the self-weight of the truck.

3.7.2 Straddle carriers (Figs 75–77)
For straddle carriers:

$$W_i = f_d\left(U_i + \frac{W_c}{M}\right) \tag{3.12}$$

where:

Fig. 75.   Three generations of straddle carriers at Europe Container Terminus, Rotterdam. The one on the left can place one container over another. The one in the centre can place a container over two others and the one on the right can place a container over three others. This evolution took place during the 1970s and early 1980s

Fig. 76.   Eight-wheel asymmetric straddle carrier handling 40 ft container

W_i = wheel load of laden plant (kg)
U_i = wheel load of unladen plant (kg)
W_c = weight of container (kg)
M = total number of wheels on plant
f_d = dynamic factor.

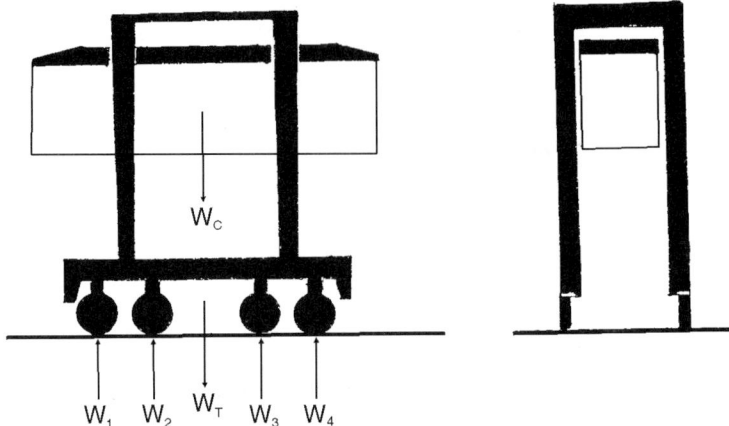

Fig. 77.   Dimensions and weights used in wheel load calculations

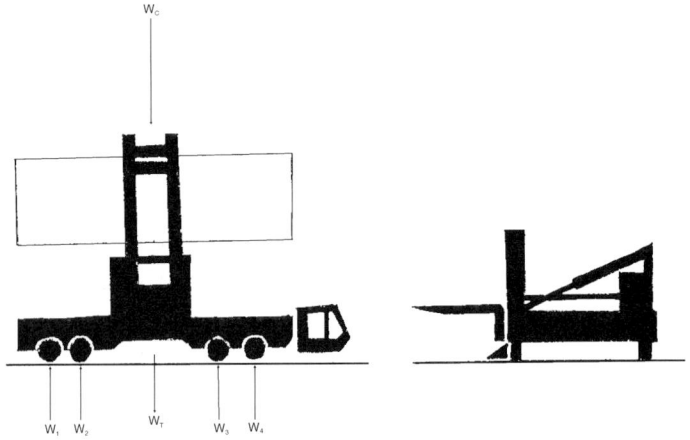

Fig. 78.   Dimensions and weights used in wheel load calculations

3.7.3   Side lift trucks (Fig. 78)

Equation (3.12) is used

$$W_i = f_d \left(U_i + \frac{W_c}{M} \right)$$

where the parameters are as defined previously.

3.7.4   Yard gantry cranes (Figs 79 and 80)

In the case of gantry cranes:

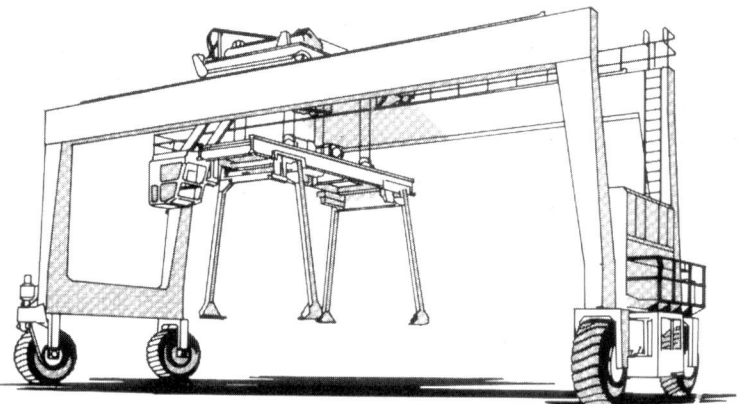

Fig. 79.   Rubber-tyred gantry crane (RTG). Individual wheel loads can exceed 50 t

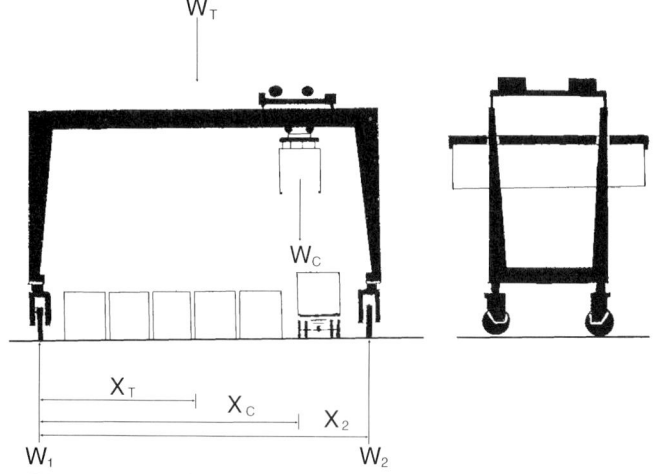

Fig. 80.   Dimensions and weights used in wheel load calculations

$$W_1 = f_d \left(U_1 + \frac{A_1 W_c}{M} \right) \qquad (3.13)$$

$$W_2 = f_d \left(U_2 + \frac{A_2 W_c}{M} \right) \qquad (3.14)$$

where:

W_1 = wheel load on side 1 (kg)
W_2 = wheel load on side 2 (kg)

W_c = weight of container (kg)
M = number of wheels on each side (possibly 2)
f_d = dynamic factor.

$A_1 = 1 - (X_c/X_2)$
$A_2 = X_c/X_2$

U_1 = unladen weight of gantry crane on each wheel of side 1 (kg)
U_2 = unladen weight of gantry crane on each wheel of side 2 (kg)
X_2 and X_c are shown in Fig. 80.

Note. The front and rear wheels may have different unladen loads. This is taken into account by using the equation for both wheels on each side with their respective f_d values.

3.7.5 Tractor and trailer systems (Figs 81 and 82)
In this example:

$$W_1 = f_d \left(U_1 + \frac{W_c(1 - A)(1 - B)}{M_1} \right) \tag{3.15}$$

$$W_2 = f_d \left(U_2 + \frac{W_c(1 - A)B}{M_2} \right) \tag{3.16}$$

$$W_3 = f_d \left(U_3 + \frac{W_c A}{M_3} \right) \tag{3.17}$$

where:

W_1 = load on front wheels of tractor (kg)
W_2 = load on rear wheels of tractor

Fig. 81. In some places, specialized off-highway tractor units are used to marshall specially developed trailers. In this case, a special small-wheeled trailer is used to reduce volume when the trailer is on board a ship

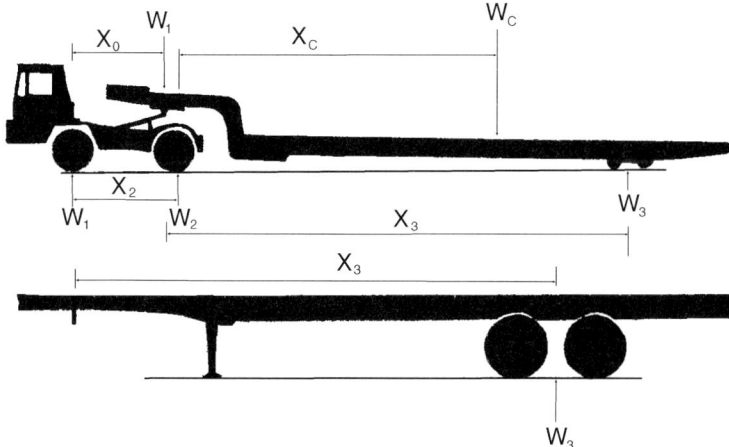

Fig. 82.   Dimensions and weights used in wheel load calculations

W_3 = load on trailer wheels (kg)
W_c = weight of container (or load) (kg)
M_1 = number of front wheels on tractor
M_2 = number of rear wheels on tractor
M_3 = number of wheels on trailer
U_1 = load on front wheels of tractor, unladen (kg)
U_2 = load on rear wheels of tractor, unladen (kg)
U_3 = load on trailer wheels, unladen (kg)
f_d = dynamic factor

$A = X_c / X_3$
$B = X_b / X_2$

X_c, X_b, X_3 and X_2, are shown in Fig. 82.

3.7.6   Mobile cranes (unladen) (Fig. 83)
For the case of a mobile crane the wheel load is:

$$W = W_T / M \qquad (3.18)$$

where W_T is the self-weight of the crane and M is the total number of wheels on the crane.

3.7.7   Materials handling equipment
Industrial hardstandings are frequently trafficked by smaller items of plant that handle pallets or individual pieces. For example, mobile transporters are required with many storage systems and they impose loads through the contact areas of their tyres. Pallet transporters (Fig. 84)

Fig. 83. Mobile cranes often use outriggers to enhance stability. This can constitute a critical load configuration

Fig. 84. Pallet transporter

are hand-pushed trailers carrying a maximum load of 20 kN on very small wheels with an average contact pressure of 9 N/mm^2. The point load design method should be used for such equipment, but in hardstandings where they operate the critical loading is often highway vehicle loading.

Counterbalanced fork lift trucks (Fig. 85) are front loading telescopic mast vehicles with an average load of 30 kN. Height is usually limited to

Fig. 85.   Counterbalance fork lift truck

8.5 m for stability reasons and the maximum wheel load occurs from the rear wheels when the truck is unladen. Whilst these trucks are designed principally to operate inside warehouses, they sometimes operate on external hardstandings.

4.   Design

4.1   Introduction

Two design procedure are presented in this chapter. The first comprises calculating stresses resulting from the loading regime and ground conditions and comparing those stresses with the strength of the concrete. This method applies to circular patch loads, i.e. those loads which can be defined in terms of a magnitude and a circular patch over which the load can be considered to be applied uniformly. The second comprises a point load design procedure which applies in the case of other categories of locally applied loading. Because hardstanding design is an exercise in serviceability, the design procedure is based upon ultimate load analysis using partial safety factors for materials and loads appropriate to serviceability. Concrete characteristic flexural strength is used as the basis of stress assessment and load induced stresses are computed using classical methods in the case of patch loading and by a finite element analysis in the case of point loading. In this latter case, a Design Chart has been developed to allow concrete slab thickness to be read directly.

The following factors have to be taken into account in hardstanding design.

(*a*) loading regime
(*b*) strength of concrete
(*c*) strength of existing ground and effect of the sub-base.

4.2   Loading regime

Loading was covered in Chapter 3 and an example of hardstanding damage caused by overloading is shown in Fig. 86. The position of the load relative to the slab edge is critical and three alternative cases may need to be considered, namely internal loading (greater than 0.5 m from edge slab), edge loading and corner loading. In the case of patch loading, three equations can be used to calculate stresses at each of the three positions. In the case of point loading, the Design Chart is based upon the

Fig. 86.   Damage sustained by this bay was caused by overloading

less onerous condition of internal loading. Pavements designed according to the Design Chart need to have full structural connectivity between neighbouring slabs. This is usually accomplished by dowel bars (see Chapter 2). In the case of internal or edge loading, the maximum stress occurs beneath the heaviest load at the underside of the slab. Corner loading creates tensile stress at the upper surface of the slab some distance away from the corner. This distance can be calculated from:

$$d = 2[(2^{0.5})\,r\,l]^{0.5} \tag{4.1}$$

where r is the radius of the loaded area (in mm), l is the radius of relative stiffness (in mm) and d is the distance from the slab corner to the position of maximum tensile stress (in mm).

4.3   Strength of concrete

Industrial hardstandings are frequently constructed from C30 or C40 concrete with a minimum cement content of $300\,\text{kg/m}^3$ and with a slump of 50 mm or less. Design is based on comparing the concrete characteristic flexural strength with the calculated flexural stresses, whereas specification is by characteristic compressive strength. Table 4.1 shows flexural strength values for a range of commonly used concretes. The table includes values for C10 concrete. Table 1.12 in Chapter 1 describes various concretes with low cement contents which are used commonly as base materials. In particular, drylean concrete and DTp Cement Bound Material Category 4 (CBM4) can both be considered as C10 concrete for design purposes and are included in the Design Chart in Fig. 94. Because

Table 4.1.   *Mean and characteristic 28 day flexural strength values for various concrete mixes*

	Flexural strength: N/mm^2	
	Mean	Characteristic
Plain C30 concrete	2.0	1.4
C30 concrete 20 kg/m^3 ZC 60/1.00 steel fibre[a]	2.8	2.0
C30 concrete 30 kg/m^3 ZC 60/1.00 steel fibre	3.2	2.2
C30 concrete 40 kg/m^3 ZC 60/1.00 steel fibre	3.8	2.7
Plain C40 Concrete	2.4	1.7
C40 concrete 20 kg/m^3 ZC 60/1.00 steel fibre	3.2	2.2
C40 concrete 30 kg/m^3 ZC 60/1.00 steel fibre	3.6	2.5
C40 concrete 40 kg/m^3 ZC 60/1.00 steel fibre	4.2	3.2
Plain C10 Concrete	0.9	0.6

[a] ZC 60/1.00 refers to a commonly used anchored bright wire fibre of total length 60 mm and wire diameter 1.00 mm

hardstandings fail by becoming unserviceable, a partial safety factor of one can be applied to characteristic strength in hardstanding design.

4.4   Strength of existing ground and effect on sub-base

The design methods require a value for the modulus of subgrade reaction K, which defines the deformability of the material beneath the hardstanding. The following four values of K are used in the design procedures:

- $K = 0.013$ N/mm^3, very poor ground
- $K = 0.027$ N/mm^3, poor ground
- $K = 0.054$ N/mm^3, good ground
- $K = 0.082$ N/mm^3, very good ground (no sub-base needed).

The beneficial effect of a granular sub-base is taken into account by increasing K according to the thickness and strength of the sub-base as shown in Table 1.4, Section 1.2.3.

4.5   Stress in concrete

The stress in an industrial hardstanding depends upon:

(a) the properties of the subgrade
(b) the loading regime, i.e. the combination of either patch loads or point loads (the design procedures do not allow a combination of both patch and point loads)

Table 4.2. Fatigue factors by which calculated stresses should be multiplied

Number of load repetitions	Fatigue factor
1	1
50	1.25
5000	1.5
50 000	1.75
4 000 000	2
25 000 000	3

(*c*) the thickness of the industrial hardstanding
(*d*) the strength of the sub-base.

In many instances, there will be a choice between the patch load design method and the point load design method. In the case of loads applied via pneumatic tyres, the patch load design method should be used. In the case of loads applied through steel or solid rubber-tyred wheels, or through container corner castings, the point load method should be used. In some cases, the designer may wish to use both methods as a check.

The fatigue effect must be considered since it is common for hardstandings to be subjected to repetitive loading. This can be achieved by increasing the calculated stresses so as to obtain the stress which would be equivalent to the multiple load stress were the load to be applied once. Table 4.2 shows the factors by which calculated stresses should be multiplied to account for fatigue. In the case of the design for point loads, the fatigue factors have been built into the Design Chart.

4.6 Design method for patch loads using modified Westergaard equations

Highway vehicles and handling equipment with pneumatic rubber tyres apply loads to the surface of a hardstanding as *patch loads* (Fig. 87). In most cases, sufficient accuracy is gained by assuming the patch is circular and applying uniform stress throughout the patch. However, these assumptions lead to minor errors — the contact patch shape in the case of a commercial vehicle is nearer a rectangle than a circle but to assume accurately shaped patch loads would preclude the use of modified Westergaard equations. The error in assuming constant stress circular loading is very small and is conservative, i.e. true analysis would lead to a slightly thinner hardstanding.

The maximum flexural tensile stress occurs at the bottom (or top in the case of corner loading) of the slab under the heaviest wheel load. The maximum stress under a patch load can be calculated by Westergaard[4.1] and Timoshenko equations as follows.

Fig. 87.   Front lift truck loads can be assessed using the patch load design method

(*a*) Patch load in mid-slab (i.e. more than 0.5 m from slab edge):

$$\sigma_{max} = \frac{0.275(1 + \nu)}{h^2} P \log\left(\frac{0.36\,Eh^3}{Kb^4}\right) \qquad (4.2)$$

(*b*) Patch load at edge of slab:

$$\sigma_{max} = 0.529(1 + 0.54\nu)\frac{P}{h^2}\log\left(\frac{0.20\,Eh^3}{Kb^4}\right) \qquad (4.3)$$

(*c*) Patch load at slab corner:

$$\sigma_{max} = \frac{3P}{h^2}\left[1 - \left(\frac{1.41b}{\{Eh^3/[12(1 - \nu^2)K]\}^{0.25}}\right)^{0.6}\right] \qquad (4.4)$$

where

σ_{max} = flexural stress (in N/mm^2)
P = point load (in N,) i.e. characteristic wheel load × fatigue factor
ν = Poisson's ratio, usually 0.15
h = slab thickness (in mm)
E = elastic modulus, usually 20 000 N/mm^2
K = Modulus of subgrade reaction (in N/mm^3)
b = radius of tyre contact zone (in mm)
        = $(P/\pi p)^{1/2}$
p = contact stress between wheel and hardstanding (in N/mm^2).

Twin wheels bolted side-by-side are assumed to be one wheel transmitting half of the axle load to the hardstanding. In certain cases wheel loads at one end of an axle magnify the stress beneath wheels at the opposite end of the axle, a distance S away (S is measured between load patch centres). To calculate the stress magnification, the characteristic length (radius of relative stiffness, L) has to be found from Equation (4.5). Note that this method of assessing the effect of wheel proximity requires a knowledge of the modulus of subgrade reaction K. In cases where only the California Bearing Ratio (CBR) is known, the proximity factor method set out in Section 3.6 should be used instead. (Alternatively, Table 1.6 in Chapter 1 can be used to obtain K value if the CBR value is known).

$$L = \left(\frac{Eh^3}{12(1 - \nu^2)K} \right)^{0.25} \qquad (4.5)$$

Once L has been evaluated, the ratio S/L can be determined so that Fig. 88 can be used to find M_t/P (M_t is the tangential moment). The stress under the heaviest wheel should be increased to account for the other wheel. This is calculated by adding an appropriate stress:

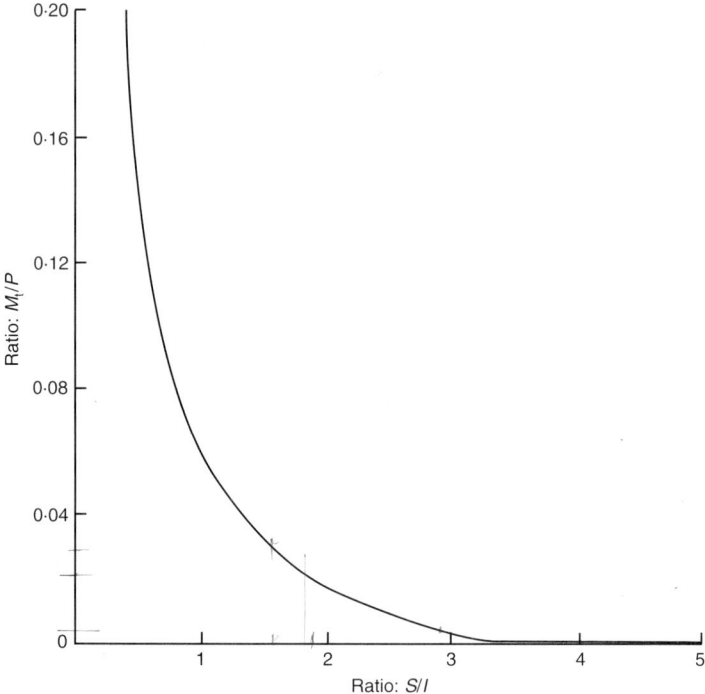

Fig. 88. Relationship between ratio S/L and ratio M_t/P used in assessing the influence of load proximity

$$\sigma_{\text{add}} = \frac{M_{\text{t}}}{P}\frac{6}{h^2}P_2 \tag{4.6}$$

where P is the greater patch load and P_2 is the other patch load.

The next step is to sum the stresses and verify that the characteristic flexural strength has not been exceeded (Table 4.1) for the prescribed concrete mix.

4.7 Highway vehicle example using Westergaard equations

Highway vehicles on industrial hardstandings often perform tight turns (Fig. 89) and an example is now worked through. Consider a highway vehicle with a rear axle load of 8000 kg and assume a slab thickness of 200 mm on good ground ($K = 0.054\,\text{N/mm}^3$). Note that this value of K could be enhanced by the provision of a sub-base according to Table 1.4 but this benefit is ignored in this example. The fatigue factor is 2.0 so the hardstanding can withstand 4 000 000 such repetitions. The design axle load is 160 000 N (i.e. 80 000 N × fatigue factor of 2.0) so the design wheel load is 80 000 N. The following parameters are known or calculated:

$p =$ contact stress between wheel and hardstanding $= 0.7\,\text{N/mm}^2$
$b =$ radius of contact area $= (W/\pi p)^{1/2}$
$\quad = (80\,000/\pi\,0.7)^{1/2}$
$\quad = 191\,\text{mm}.$

Substituting known values into Equation (4.4) yields:

Fig. 89. Highway vehicles typically execute severe turning movements on industrial hardstandings

$$\sigma_{max} = 0.529(1 + 0.54 \times 0.15)\frac{80\,000}{200^2}\log\left(0.2\frac{20\,000 \times 200^3}{0.054 \times 191^4}\right)$$

$$\sigma_{max} = 3.03\,\text{N/mm}^2 \text{ beneath one wheel.}$$

Assuming the wheel at the other end of the axle to be 2.7 m away ($S = 2700$ mm), the radius of relative stiffness (L), using Equation (4.5) is:

$$L = \left(\frac{20\,000 \times 200^3}{12(1 - 0.15^2)0.054}\right)^{0.25}$$

$$L = 709\,\text{mm}$$

Thus

$$S/L = 2700/709 = 3.8$$

From Fig. 88

$$M_t/P = 0.005$$

Therefore, using Equation (4.6)

$$\sigma_{add} = 0.005\left(\frac{6}{200^2}\right)80\,000$$

$$= 0.06\,\text{N/mm}^2$$

The total stress is thus $3.03 + 0.06\,\text{N/mm}^2 = 3.09\,\text{N/mm}^2$.

By comparing this stress with the characteristic strength of C40 concrete reinforced with 40 kg/m^3 steel fibres (3.2 N/mm^2), it can be seen that the proposed mix is satisfactory. The inclusion of a 250 mm thick granular sub-base would enhance K from 0.054 to 0.073 which would reduce the stress further below the characteristic value.

Figures 90 to 93 show some illustrations of hardstanding failure and success.

4.8   Design method for point loads using Design Chart

The above example illustrates the complexity of undertaking calculations in the case of patch loads. In the case of point loads, in order to eliminate much of the effort, a Design Chart has been developed using finite element analysis. The design chart in Fig. 94 assumes that there is full transfer of load between neighbouring slabs so no account is taken of different load positions. Figure 94 can be used as follows:

(1) Assess the existing conditions.

Determine the Actual Point Load (APL) and modulus of subgrade reaction K values from Chapter 3 to confirm the category of subgrade.

Fig. 90.   The corner of this bay has failed as a result of overloading. The mesh reinforcement has held the pieces together in a pattern which reflects the orientation of the wires

Fig. 91.   This mid-bay crack has occurred as a result of excessive joint spacing. It developed as a result of the action of heavy container handling plant

Note that if only CBR values are known, Table 1.6 provides corresponding K values. Note that the Design Chart assumes that the slab is constructed over material with an effective CBR of 5%. In the case of subgrades with a CBR of 5% or greater, a nominal sub-base 150 mm thick can be provided. In the case of subgrades with a CBR of 1%, 2%, 3% or 4% (subgrade CBR values should be quoted to whole numbers), there is a choice between using additional sub-base material (see Section

Fig. 92. Corner cracking has occurred in each of the four bays as a result of overloading

Fig. 93. The reinforced concrete bays are withstanding the stresses applied by these steel jockey wheels. Compare this with Fig. 71 where similar wheels indented the bituminous surfacing material

1.2) or continuing to use 150 mm thickness of sub-base but with the addition of capping material between the sub-base and the subgrade. Capping material is usually locally available low-cost material such as selected hardcore, crushed concrete, crushed rock not complying with the requirements of Type 2 materials or a mixture of clay and gravel (frequently specified and known as 'hogging' in the south of England). The capping material is required to have a CBR of 15%.

If additional sub-base material is used without capping material, the sub-base thickness should be:

- 5% CBR subgrade 150 mm thick sub-base
- 4% CBR subgrade 250 mm thick sub-base

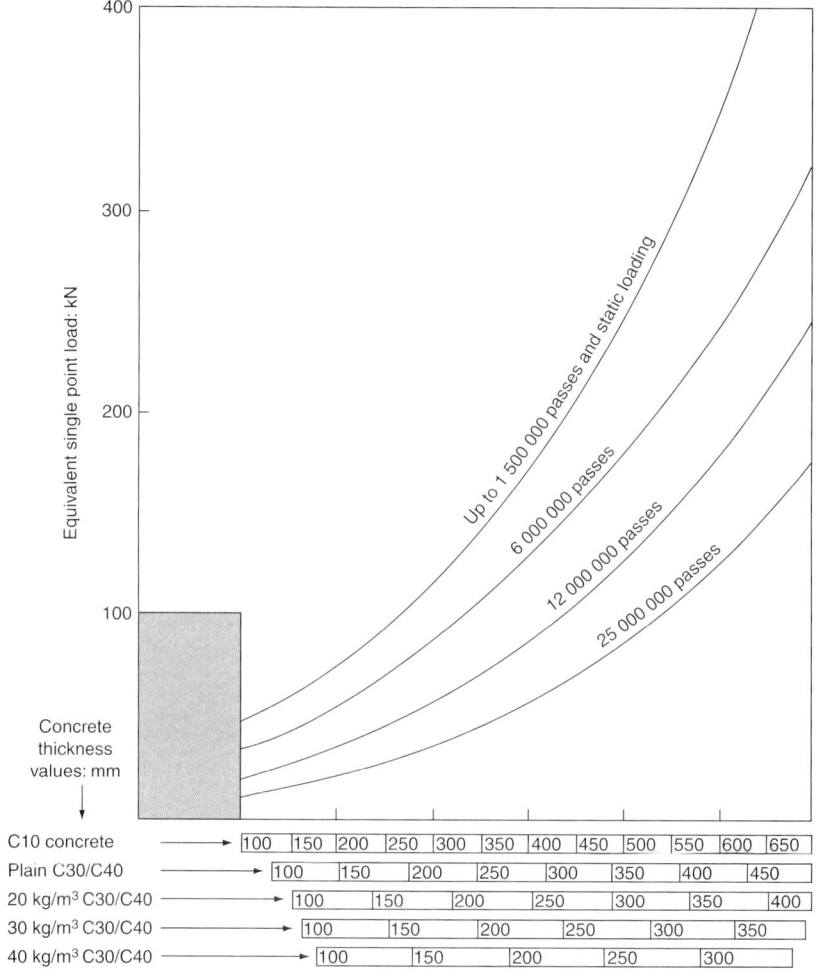

Fig. 94.   Point load design chart

- 3% CBR subgrade   400 mm thick sub-base
- 2% CBR subgrade   600 mm thick sub-base
- 1% CBR subgrade   1000 mm thick sub-base.

If 150 mm thick sub-base is to be provided and capping is to be installed beneath the sub-base, the capping thickness should be:

- 5% CBR subgrade   zero thickness capping below 150 mm thick sub-base
- 4% CBR subgrade   250 mm thickness capping below 150 mm thick sub-base

- 3% CBR subgrade   400 mm capping below 150 mm thick sub-base
- 2% CBR subgrade   600 mm thick capping below 150 mm thick sub-base
- 1% CBR subgrade   1000 mm thick capping below 150 mm thick sub-base.

(2) Calculate the additional stress generated by point loads or wheels in close proximity.

(a) If the distance between loads (S) is greater than 3 m, the APL can be used directly (depending on the radius of contact zone) to calculate the thickness of the slab using the Design Chart. In this case, go directly to Stage 6.

(b) If the distance between loads is less than 3 m, the radius of relative stiffness (L) has to be determined. Table 4.3 shows values of radius of relative stiffness for different K values and slab thicknesses; E = elastic modulus = $20\,000\,\text{N/mm}^2$; ν = Poisson's ratio = 0.15.

Table 4.3.   Radius of relative stiffness values (l) for different slab thicknesses and support conditions

Slab thickness: mm	Modulus of subgrade reaction K: N/mm^3			
	0.013	0.027	0.054	0.082
150	816	679	571	515
175	916	763	641	578
200	1012	843	709	639
225	1106	921	774	698
250	1196	997	838	755
275	1285	1071	900	811
300	1372	1143	961	865

From Fig. 88, determine M_t/P, the ratio of the tangential moment to the greater point load by calculating S/L. Then use Equation (4.6) to determine the stress to add where P is the greater point load (in N) and P_2 is the other point load (in N).

(3) From Table 4.1, select a proposed concrete mix and hence characteristic strength, σ_{char}.

(4) When two point loads are acting in close proximity (i.e. less than 3 m apart), the greater point load (P) produces a flexural strength σ_{max} directly beneath the point of application. The nearby smaller point load (P_2) produces additional stress σ_{add} beneath the larger load. Calculate σ_{max} from:

$$\sigma_{max} = \sigma_{char} - \sigma_{add} \tag{4.7}$$

(5) Calculate the Single Point Load (SPL) which, acting alone, would generate the same flexural stress as the actual loading configuration from

$$SPL = APL(\sigma_{char}/\sigma_{max}) \tag{4.8}$$

where APL is the Actual Point Load.

(6) Prior to using the Design Chart, it is necessary to modify the SPL to account for contact area as well as wheel proximity to obtain the ESPL (Equivalent Single Point Load). The Design Chart applies directly when point loads have an effective radius of contact between 150 mm and 250 mm. Some smaller rubber-tyred fork lifts have a contact radius of less than 150 mm and some vehicles have a contact radius greater than 250 mm. In these cases multiply the point load by the factor listed in Table 4.4 before using the Design Chart.

(7) Use the Design Chart for the mix selected in (3) to determine slab thickness and return to (3) if an alternative concrete mix is required.

Table 4.4.   *Point load multiplication factors for different values of K and for loads with a radius of contact outside the range 150 mm to 250 mm*

Radius of contact: mm	Modulus of subgrade reaction K: N/mm^3			
	0.013	0.027	0.054	0.082
50	1.5	1.6	1.7	1.7
100	1.2	1.2	1.3	1.3
150	1.0	1.0	1.0	1.0
200	1.0	1.0	1.0	1.0
250	1.0	1.0	1.0	1.0
300	0.9	0.9	0.9	0.9

4.9 Design example for multiple point loading using Design Chart

Consider a concrete hardstanding subjected to two point loads. A 60 kN point load is applied 1 m away from a 50 kN point load. The 60 kN point load has a contact zone radius of 100 mm and the 50 kN point load has a 300 mm radius. The existing ground conditions are poor ($K = 0.027$ N/mm^3). Assume 1 500 000 repetitions of the load.

Assume thickness of slab $= 225\,\text{mm}$ — this assumption is required so that Table 4.3 can be used to assess the radius of relative stiffness, L. The radius of relative stiffness, L, is given as $921\,\text{mm}$ by Table 4.3. The distance between point loads $= 1\,\text{m}$ (i.e. $S = 1000\,\text{mm}$). Thus,

$$S/L = 1000/921 = 1.086$$

From Fig. 88

$$M_t/P = 0.053$$

so from Equation (4.6),

$$\sigma_{\text{add}} = 0.053\left(\frac{6}{200^2}\right)50\,000$$

$$\sigma_{\text{add}} = 0.4\,\text{N/mm}^2$$

Try steel fibre reinforced C30 concrete with a characteristic strength of $2.2\,\text{N/mm}^2$ ($30\,\text{kg/m}^3$ steel fibre dosage (see Table 4.1)), i.e. $\sigma_{\text{char}} = 2.2$ N/mm^2

$$\sigma_{\text{max}} = \sigma_{\text{char}} - \sigma_{\text{add}}$$

$$= 2.2 - 0.4$$

$$= 1.8\,\text{N/mm}^2$$

This is the maximum flexural stress which the 60 kN load can be allowed to develop. The SPL calculated using Equation (4.8) is thus:

$$= 60\,(2.2/1.8) = 73\,\text{kN}$$

From Table 4.4, the modified factor to be applied to the SPL to obtain the ESPL is 1.2, giving

$$73 \times 2 = 88\,\text{kN}$$

From the Design Chart (Fig. 94) the thickness of slab required is 225 mm.
   A 175 mm thick C30 concrete slab incorporating $20\,\text{kg/m}^3$ steel fibre is thus adequate for this design. The original assumption of 225 mm was conservative and this design is satisfactory.

4.10   Eight-wheel straddle carrier design examples
In this example a straddle carrier operation is assessed for loading and subsequent use with the Design Chart to produce a pavement section. In the loading calculations, the damaging effect of one side of the item of plant is considered, as explained in this example.

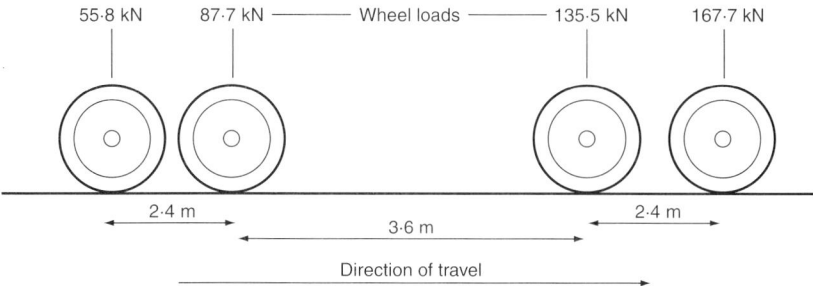

Fig. 95.   Straddle carrier wheel loads during braking

4.10.1   Data

Unladen weight of straddle
carrier including spreader beam $=$ 56 310kg (W_T)
Critical container weight $=$ 22 000kg
Track width $=$ 4.5m
Wheel spacings $=$ 2.4 m–3.6 m–2.4 m (see Fig. 95)
Number of likely passes of straddle
carriers over the most highly
trafficked part of the pavement
during design life of pavement $=$ 960 000
CBR of soil $=$ 5%
Sub-base $=$ 225 mm

4.10.2   Example I

Having defined the foundation material properties, the base material is
now calculated: it is dependent on the load applied.

Total number of wheels on plant $= 8$
Wheel load of unladen plant (kg) $= 56\,310/8 = 7039\,kg$
Weight of critical container (kg) $= 22\,000\,kg$, see Chapter 3, Section
3.1
$f_d =$ dynamic factor for braking $= \pm 50\%$ for extreme wheels, see next
paragraph for inner wheels
Static wheel load $= 7039 + 22\,000/8 = 9789\,kg = 97.9\,kN.$

The proximity of the wheel loads is now considered to assess their
stress interaction using the method shown in Chapter 3, Section 3.6 to
calculate the effective depth.

$$\text{Effective depth} = 300\sqrt[3]{\frac{35\,000}{5 \times 10}} = 2664\,\text{mm}$$

From Table 3.4, the proximity factor can be interpolated to be 1.14. Therefore the Effective Static Wheel Load is $97.9 \times 1.14 = 111.6\,\text{kN}$. Consider the most adverse loading case of braking and apply the appropriate dynamic factor $\pm 50\%$ to the wheels at the extreme front and rear, applying the increase in load to the front wheels and the decrease to the rear wheels. The inner wheel loads need to be similarly adjusted, but by using a factor lower than $\pm 50\%$ determined by considering their relative distance from the vehicle's centre line. In this case, each extreme wheel is 4.2 m from the centre of the vehicle and each inner wheel is 1.8 m from the centre. Therefore, the lower braking factor to be applied to the inner wheels is $\pm 21.4\%$, i.e. ($\pm 50\% \times 1.8/4.2$). We now need to express the four load values which will pass over one spot as an equivalent number of passes of the highest wheel load (167.7 kN) as follows. The damaging effect equation in Section 3.1 is applied to each wheel load in turn.

- **Front** wheel is equivalent to one pass of a load of 167.7 kN.
- **Second** wheel is equivalent to $(135.5/167.7)^{3.75}$, i.e. 0.45 equivalent passes of the front wheel load.
- **Third** wheel is equivalent to $(87.7/167.7)^{3.75}$, i.e. 0.09 equivalent passes of the front wheel load.
- **Fourth** wheel is equivalent to $(55.8/167.7)^{3.75}$, i.e. 0.02 equivalent passes of the front wheel load.

All of the repetitions are converted to an equivalent number of repetitions of the heaviest wheel so that the Equivalent Single Wheel Load used in the Design Chart is derived from the heaviest wheel load. It would be unsafe to convert wheel loads to one of the plant's lower wheel load values. Therefore, each time the straddle carrier passes over one spot, it applies the equivalent of $(1 + 0.45 + 0.09 + 0.02) = 1.56$ repetitions of the front wheel load of 167.7 kN. This means that the pavement needs to be designed to accomodate 1.5 million passes (i.e. $1.56 \times 960\,000$) of a load of 167.7 kN. The base thickness Design Chart can now be used as follows.

- On the vertical axis, the Equivalent Single Wheel Load is 167.7 kN.
- The appropriate curve is the one corresponding to 1.5 million passes.
- The following alternative thicknesses can be used:

○ C10 concrete   400 mm
○ Plain C30/C40 concrete   300 mm
○ 20 kg/m³ steel fibre C30/C40 concrete   250 mm
○ 30 kg/m³ steel fibre C30/C40 concrete   225 mm
○ 40 kg/m³ steel fibre C30/C40 concrete   215 mm.

4.10.3   Example 2

Consider how the pavement section required would change if alternative dynamic factors were used. For example, if the straddle carriers were to brake whilst cornering, the wheel loads would increase by 60% of their static value (i.e. $0.6 \times 111.6 = 67.0$ kN) so that the wheel loads would be as shown in Fig. 96.

We now need to express the four load values which will pass over one spot as an equivalent number of passes of the highest wheel load (224.7 kN) as follows. The damaging effect equation in Section 3.1 is applied to each wheel load in turn.

- **Front** wheel is equivalent to one pass of a load of 224.7 kN.
- **Second** wheel is equivalent to $(202.5/224.7)^{3.75}$, i.e. 0.68 equivalent passes of the front wheel load.
- **Third** wheel is equivalent to $(154.7/224.7)^{3.75}$, i.e. 0.25 equivalent passes of the front wheel load.
- **Fourth** wheel is equivalent to $(122.8/224.7)^{3.75}$, i.e. 0.10 equivalent passes of the front wheel load.

Therefore, each time the straddle carrier passes over one spot, the outside wheels apply the equivalent of $(1 + 0.68 + 0.25 + 0.10) = 2.03$ repetitions of the front wheel load of 224.7 kN. This means that the pavement needs to be designed to accommodate 2 million passes (i.e. $2.03 \times 960\,000$) of a load of 224.7 kN. The base thickness Design Chart can now be used as follows:

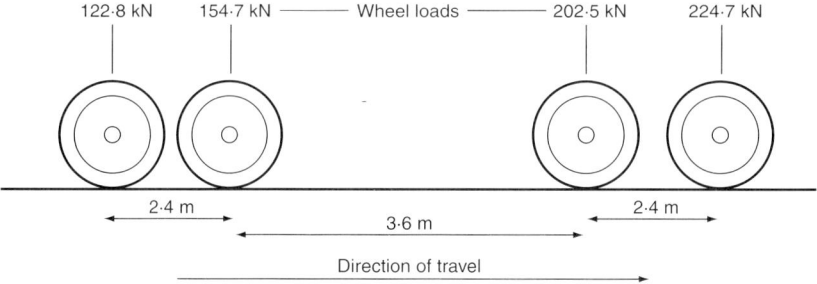

Fig. 96.   Straddle carrier wheel loads during braking and cornering

- On the vertical axis, the Equivalent Single Load is 224.7 kN.
- A 2 000 000 pass curve has to be interpolated between the 1 500 000 and the 6 000 000 curves.
- The following alternative thicknesses can be used:

   - C10 concrete   500 mm
   - Plain C30/C40 concrete   350 mm
   - 20 kg/m^3 steel fibre C30/C40 concrete   325 mm
   - 30 kg/m^3 steel fibre C30/C40 concrete   275 mm
   - 40 kg/m^3 steel fibre C30/C40 concrete   250 mm.

4.10.4   Example 3

Finally, consider the case where straddle carriers are running freely on a smooth surface so that no dynamic factors need be applied. In this configuration, the wheel loads are as in Fig. 97.

The pavement withstands four repetitions of a wheel load of 111.6 kN as each straddle carrier wheel passes so the pavement must be designed to withstand 3 840 000 passes (say 4 000 000) of an Equivalent Single Load of 111.6 kN. The following alternative thicknesses can be used:

- C10 concrete   325 mm
- Plain C30/C40 concrete   250 mm
- 20 kg/m^3 steel fibre C30/C40 concrete   225 mm
- 30 kg/m^3 steel fibre C30/C40 concrete   200 mm
- 40 kg/m^3 steel fibre C30/C40 concrete   175 mm

In the case of plain concrete, different operational conditions led to the pavement thicknesses required varying between 250 mm and 350 mm. In some cases it may be possible to take advantage of known modes of operation and proportion the pavement courses to meet the thicknesses required exactly. Whilst this may reduce initial construction costs, it has the disadvantage of constraining future operations and may lead to

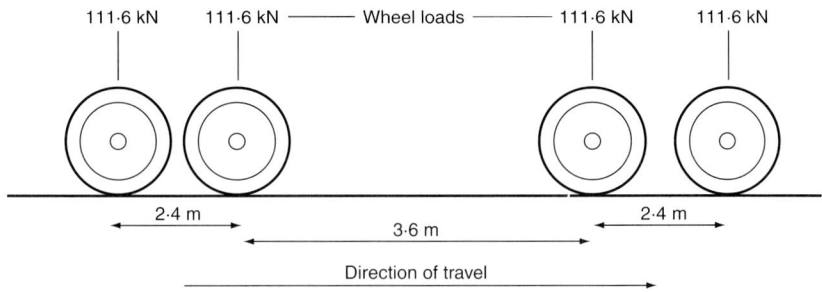

Fig. 97.   Straddle carrier wheel loads during free running

Table 4.5. Multiplication factors for dynamic wheel loads

Condition	Plant type	Multiplication factor
Braking	Front lift truck	1.3
	Straddle carrier	1.5
	Side lift truck	1.2
	Tractor and trailer	1.1
Cornering	Front lift truck	1.4
	Straddle carrier	1.6
	Side lift truck	1.3
	Tractor and trailer	1.3
Acceleration	Front lift truck	1.1
	Straddle carrier	1.1
	Side lift truck	1.1
	Tractor and trailer	1.1
Uneven surface	Front lift truck	1.2
	Straddle carrier	1.2
	Side lift truck	1.2
	Tractor and trailer	1.2

additional complexity in the construction process. It may prove cost effective to provide an initial pavement which will not sustain all potential operational situations and to allow the plant to become the proof testing system so that small areas may have to be strengthened later. Whilst this staged approach has the advantage of lowering initial costs, this must be balanced against the disadvantage of disruption which may occur should the pavement need to be upgraded later.

4.11   Initial slab thickness design

The following simplified design procedure can be used to establish an initial estimate of required slab thickness. The detailed procedures set out in this chapter should be followed to obtain an accurate design solution. It is assumed that the design load will not be applied to one spot in the hardstanding more than 1 500 000 times.

(1) Assess static wheel load of plant.
(2) Modify wheel load for dynamic effects according to Table 4.5.
(3) Modify wheel loads when wheels are 2.4 m apart or closer according to Table 4.6.
(4) Select plain C30/C40 concrete slab thickness required from Table 4.7.

Table 4.6. Wheel load multiplication factors

Wheel spacing: mm	Multiplication factor
300	2
600	1.9
900	1.8
1200	1.7
1800	1.5
2400	1.3

Table 4.7. Slab thickness required for various design loads

Design load: kN	Slab thickness: mm
50	150
100	200
150	250
200	300
250	350
300	400
350	425
400	450
450	475
500	500

5. Case studies and data

5.1 Thermal and moisture-related stresses in concrete slabs

The basic premise underlying most concrete pavement design methods is that stresses developed as a result of a change in temperature of moisture content of the concrete slab are contained by the provision of stress-relieving joints, whereas stresses occurring as a result of traffic or other applied loads are controlled by proportioning the thickness of the slab and its underlying supporting courses. The exception is in continuously reinforced concrete pavements (CRCPs) in which temperature and moisture-loss stresses are contained by the composite action of the reinforcement and the concrete.

Whether temperature or moisture-loss stresses are predominant depends upon many factors that are difficult to calculate. Moisture-related stresses are potentially greater than temperature-related ones by an order of magnitude. However, it is often the case that a highway pavement retains much of its moisture throughout its life. Also, moisture-related stresses develop slowly so creep often reduces them significantly. Thermal stresses, on the other hand, are often at their most severe immediately following construction as the setting concrete cools. Furthermore, temperature-related stresses are usually diurnal so creep has little mitigating effect. For this reason, temperature-related effects are of most concern in most concrete highway pavement projects.

Although moisture-loss effects are usually less important than temperature-related effects,[5.1] they need careful consideration in the case of highways constructed in a dry climate. Both temperature and moisture can cause a slab to shrink uniformly, to curl upwards at its perimeter or to curl downwards at its perimeter. The way in which these three conditions impart stress into the slab is now considered.

5.1.1 Uniform shrinkage

As a result of uniform temperature fall or moisture loss, a concrete slab will shrink uniformly about its centre on plan. Theoretically the centre will remain stationary. At a distance from the centre the slab will attempt to displace horizontally and this displacement will increase uniformly towards the edge of the slab. Frictional restraint between the underside of the concrete slab and the surface of the sub-base will inhibit or prevent this movement and so generate tensile stress within the slab. This stress in the concrete resulting from frictional restraint to shrinkage can be calculated. The force required to overcome the friction force is given by the expression, $F_f - w\mu$, where F_f is the friction force, w is the weight of the concrete (calculated by assuming a density of $24\,kN/m^3$) and μ is the coefficient of friction. As the weight of concrete generating frictional restraint increases with distance from the slab edge, the stress gradually increases to a maximum at the slab centre (there is zero stress at the edge of the slab or at the joints). Assuming the values for the coefficient of friction between concrete and sub-base and polyethylene are 0.65 and 0.15 respectively, the theoretical stresses which result from uniform shrinkage friction are shown in Fig. 98. Even without a slip membrane, the stresses are low, attaining a value of less than $0.05\,N/mm^2$ in a 6 m long slab. Figure 98 illustrates why the provision of a slip membrane is not a crucial issue. Indeed, in an unreinforced concrete pavement with

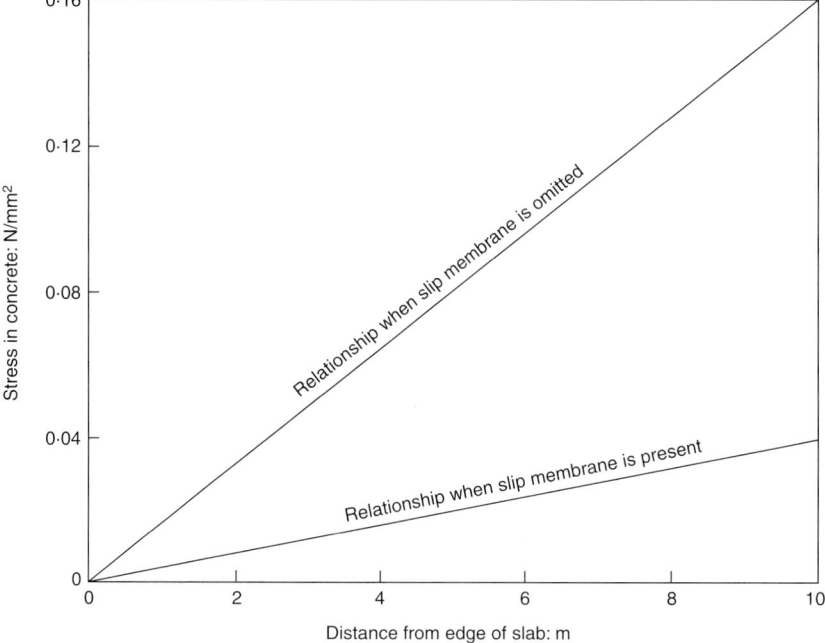

Fig. 98. Relationship between friction-induced stress developed in slab against distance from edge of slab for slabs with and without slip membranes

closely spaced joints at say 5 m spacings, the provision of a slip membrane may be detrimental in that it may concentrate movement at one joint, which can then become a maintenance problem, while other joints never operate. (It may be preferable to provide a layer of polyethylene to prevent concrete water loss into the sub-base).

5.1.2   Slab perimeter curling downwards (hogging)

As a result of the underside of the concrete slab cooling or drying faster than the top, non-uniform shrinkage develops throughout the slab with the lower concrete shrinking more than the upper. The result of this non-uniform shrinkage will be hogging of the slab. Assuming the hogging slab can be represented by a simply supported beam of length L with a uniformly distributed load equal to the concrete weight then the maximum moment $M = wL^2/8$, where L is the length of the slab (the distance between joints) and w is the dead weight of the concrete assumed to be 24 kN/m^3. The stress is calculated from the equation $\sigma/y = M/I$ where σ is the stress, y is the depth to the neutral axis, M is the bending moment and I is the second moment of area. In the extreme case, L would be the distance between joints allowing rotational freedom. Figure 99 shows stresses which would develop in the hogging situation for a 200 mm thick slab.

Figure 99 demonstrates that such behaviour would crack each bay. In fact, a temperature fall or moisture loss is usually insufficient to cause the slab to separate from its sub-base so this extreme condition rarely occurs.

5.1.3   Slab perimeter curling upwards (curling)

The result of the upper side of the slab cooling or drying faster than the underside will be that the slab will attempt to curl upwards at its edges. This curling can be represented by a cantilever of length L equal to the curled length. Assuming this cantilever carries a uniformly distributed load generated by the weight of the concrete (based upon an assumed density of 24 kN/m^3), the bending moment at the slab's point of contact with the ground is given by the expression $M = wL^2/2$. As in the case of hogging, the stress can then be calculated from the expression $\sigma/y = M/I$. Figure 100 shows curling stresses calculated for a 200 mm thick slab.

5.1.4   Calculation of slab temperature changes

In order to determine temperature-related stresses through the depth of a slab, the temperature profile at the time of set needs to be known. From that time forward, whenever the 'at set' temperature profile is replicated, temperature stresses will disappear. The at set temperature profile has

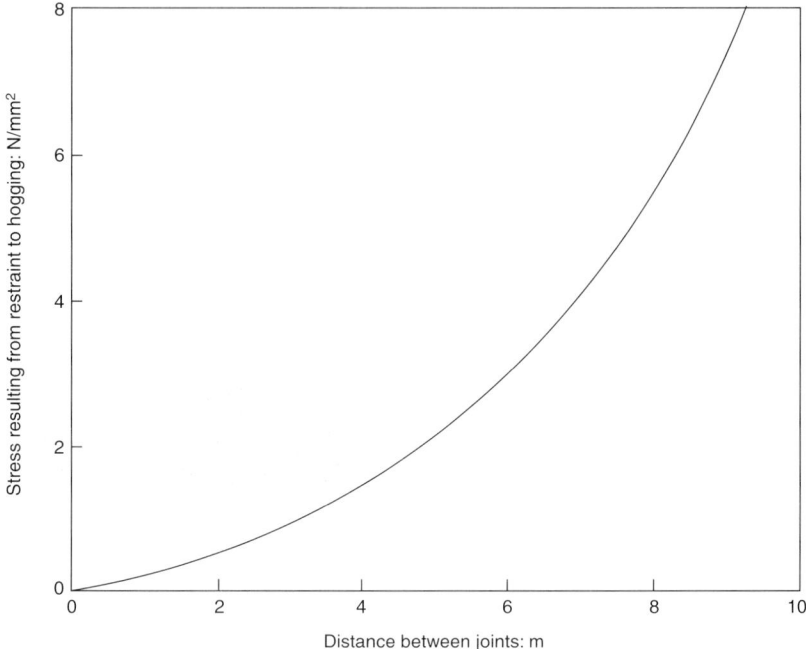

Fig. 99. Relationship between joint spacing and stress developed as a result of restraint to hogging

been investigated by several researchers who concluded that there is no standard profile of value to the designer. The profile depends upon the type of concrete, the curing regime, the weather during concreting and the time of day at which the concrete set.

Figure 101 illustrates differing temperature profiles at set and at subsequent times in different climatic conditions. Concrete that sets during a warm mid-afternoon may have a locked-in profile as shown in Fig. 101(*b*). If this slab were produced in a warm climate, then it might subsequently be subjected to a profile as in Fig. 101(*e*) during the night. In such a case, the temperature of the slab surface has fallen by, say, 20°C and the temperature of the underside of the slab has increased by 17°C. This will cause the slab to attempt to curl upwards at its perimeter (curling). The self-weight of the slab together with applied loading will attempt to keep the slab in contact with its sub-base so tensile stresses will develop at and near the upper surface of the slab. The value of these stresses will depend on the relative elastic properties of the slab and the underlying sub-base material.

The opposite effect would occur for slabs which developed their initial set during a cold morning following a relatively warm night, in which

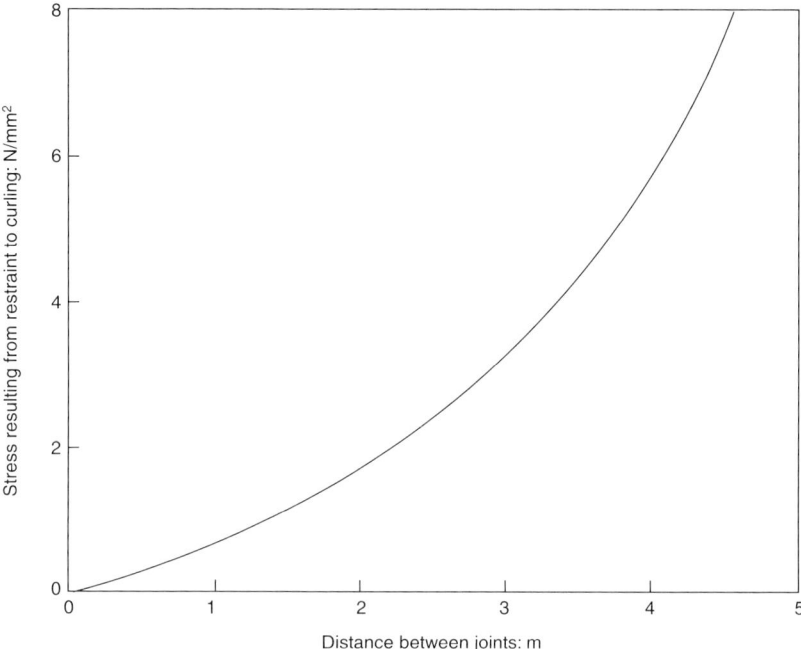

Fig. 100. Relationship between joint spacing and stress developed as a result of restraint to curling

case the at set profile shown in Fig. 101(*a*) would apply. If, subsequently, this slab were subjected to the temperature profile as shown in Fig. 101(*d*), the surface temperature rises by, say, 25°C and the underside temperature falls by 15°C. This causes the slab to attempt to curl downwards at its perimeter (hogging).

The above two cases represent extremes that slabs might be expected to endure in normal situations. In the first case, the upwards perimeter curling is caused by a temperature differential of 37°C and in the second case, the downwards perimeter curling temperature differential is 40°C. In order to gauge the magnitude of the stresses which might be generated, consider the extreme case in which the slab is fully restrained against hogging. The concrete might have a coefficient of thermal expansion of $0.000009°C^{-1}$.[5.2] Therefore, the maximum tensile strain would be 40 times 0.000009 or 0.00036. Concrete with a Young's modulus of $20\,000\,N/mm^2$ would develop a tensile stress of approximately $7\,N/mm^2$ which is sufficient to crack the concrete. This value is never attained in the concrete because full restraint is never achieved and curing regimes ensure that the 'as set' profiles shown in Fig. 101 are not actually attained in well controlled projects. In practice, it is found that the provision of transverse warping joints (i.e. joints

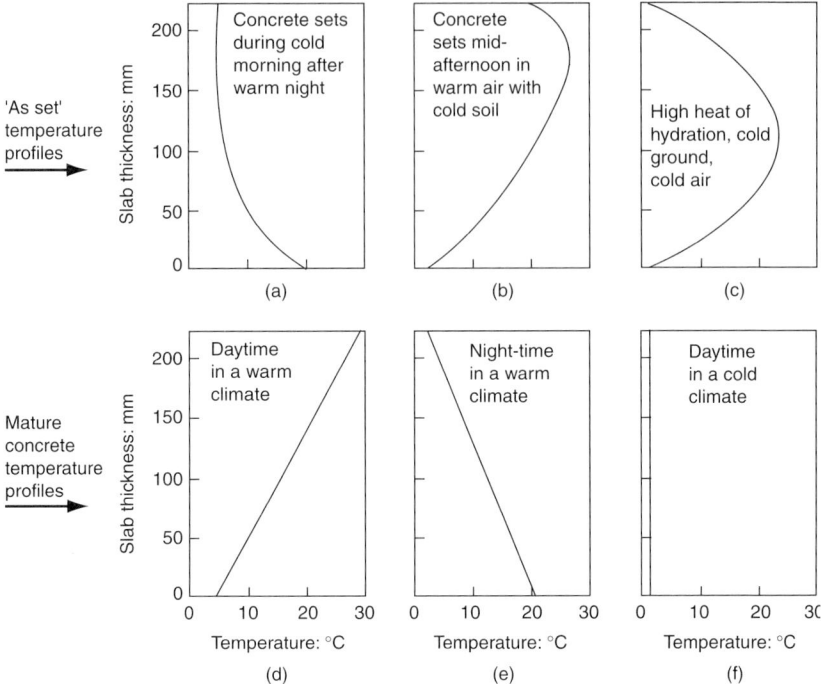

Fig. 101. Concrete slab temperature profiles at set and at subsequent times

which permit rotation at spacings of between 5 m and 30 m depending upon the level of reinforcement) is sufficient to control temperature stresses. Table 2.1 (reproduced on page 126) shows joint spacings which have been found to be sufficient to control temperature and moisture-related cracking.

Difficulties can occur when different concretes are used in a two-layer construction. The use of low heat concrete below pavement-quality concrete can lead to excessive tensile stresses — difficulties were experienced, for example, on the London Orbital Motorway (M25) which suffered transverse cracking.

Moisture-related shrinkage occurs as the concrete slab loses its free water through evaporation. Typically, pavement-quality concrete will have a water/cement ratio of 0.35. Approximately two thirds of this water is combined chemically with the cement during hydration and the remainder acts as a lubricant, creating workability. All of this free water has the opportunity to evaporate; the volume which it occupied previously is no longer present so shrinkage occurs. There may be 50 l/m^3 of water evaporating over a period of several months following construction. This represents 5% of the volume of the concrete so significant shrinkage is possible. In fact, creep and precipitation reduce the effect of shrinkage to a controllable level and the joint spacings discussed earlier are usually sufficient to prevent distress.

5.1.5   Conclusion

Theoretical stresses induced by friction are negligible and a slip membrane is not required from this point of view. However, in cases of hogging and curling, the theoretical stresses can exceed the flexural strength of the concrete. Hogging is uncommon because slip membranes reduce loss of moisture from the bottom of the slab, so preventing it from curling. Good curing of the concrete will reduce drying shrinkage and, provided the stresses are developed slowly, concrete creep will reduce curling stresses to levels which require no additional reinforcement.

5.2   Design case study
5.2.1   Details of case study

The following example illustrates the design process on a hardstanding constructed using a laser-guided screeding machine. Figure 63 (section 2.10) shows a plan of an industrial hardstanding for which a ground bearing slab design is required. The hardstanding is to be installed by a laser-guided screeding machine as one pour. The hardstanding is to be constructed over ground with a modulus of subgrade reaction K of $0.027 \, \text{N/mm}^3$. The hardstanding area of $2880 \, \text{m}^2$ is within the daily capacity of a laser-guided screeding machine, so it will be unneccesary to provide construction joints.

The hardstanding is to comprise a steel fibre reinforced concrete slab constructed over a granular sub-base. The joint spacings, slab thickness, slab strength, fibre dosage and sub-base thickness are determined as follows.

The first consideration is joint spacing. The only joints within the body of the hardstanding will be induced joints formed by saw cutting 50 mm deep slots into the slab as soon as the concrete has gained sufficient strength to permit sawing without damaging the concrete. Table 2.1 illustrated that all commonly specified concretes are suitable.

A $20 \, \text{kg/m}^3$ steel fibre C40 concrete will now be investigated. Table 4.1, reproduced here shows that this fibre reinforced concrete has a characteristic flexural strength of $2.2 \, \text{N/mm}^2$.

The slab is loaded by point loads from container corner castings and by vehicles applying patch loads. In order to account for fatigue, the two categories of loading are to be increased by the following factors.

- Container corner casting loads: $44 \times 1.5 = 66 \, \text{kN}$.
- Vehicle wheel load: $45 \times 2 = 90 \, \text{kN}$.

The above two factors (1.5 and 2) are based upon conventional fatigue relationships. Table 5.1 shows how increasing the load by various factors

Table 2.1.   Common concretes and suggested joint spacings

Concrete type	Joint spacing: m
Plain C30 concrete	6
20 kg/m^3 ZC 60/1.00 steel fibre reinforcement C30 concrete	7
30 kg/m^3 ZC 60/1.00 steel fibre reinforcement C30 concrete	8
40 kg/m^3 ZC 60/1.00 steel fibre reinforcement C30 concrete	10
Plain C40 concrete	6
20 kg/m^3 ZC 60/1.00 steel fibre reinforcement C40 concrete	8
20 kg/m^3 ZC 60/1.00 steel fibre reinforcement C40 concrete	10
20 kg/m^3 ZC 60/1.00 steel fibre reinforcement C40 concrete	12

Table 4.1.   Mean and characteristic 28 day flexural strength values for various concrete mixes

	Flexural strength: N/mm^2	
	Mean	Characteristic
Plain C30 concrete	2.0	1.4
C30 concrete 20 kg/m^3 ZC 60/1.00 steel fibre[a]	2.8	2.0
C30 concrete 30 kg/m^3 ZC 60/1.00 steel fibre	3.2	2.2
C30 concrete 40 kg/m^3 ZC 60/1.00 steel fibre	3.8	2.7
Plain C40 Concrete	2.4	1.7
C40 concrete 20 kg/m^3 ZC 60/1.00 steel fibre	3.2	2.2
C40 concrete 30 kg/m^3 ZC 60/1.00 steel fibre	3.6	2.5
C40 concrete 40 kg/m^3 ZC 60/1.00 steel fibre	4.2	3.2
Plain C10 Concrete	0.9	0.6

[a] ZC 60/1.00 refers to a commonly used anchored bright wire fibre of total length 60 mm and wire diameter 1.00 mm

will extend the life of the hardstanding. The system presented here allows different categories of loads to have different fatigue factors.

The assessment of point loads applied by container corner casting feet is to be undertaken by using the point load Design Chart shown in Fig. 94. The patch loads applied by the vehicle are to be assessed by the modified Westergaard equation method.

Table 5.1 Relationship between number of load repetitions and fatigue factor. By multiplying the actual load by the fatigue factor, a higher load is produced which can be used to ensure a hardstanding of the specified life

Number of load repetitions	Fatigue factor
1	1
50	1.25
5000	1.5
50 000	1.75
4 000 000	2
25 000 000	3

5.2.2 Assessment of container casting point loads using Design Chart

In the case of container corner casting loads, the effect of load proximity needs to be taken into account. (It is assumed that the vehicle tyres are sufficiently well separated for proximity to be ignored — this is not always the case.) Figure 88 (Section 4.6) is used to determine the ratio M_t/P. To do this, first, calculate the radius of relative stiffness L from Equation (4.5):

$$L = \left(\frac{Eh^3}{12(1 - \nu^2)K} \right)^{0.25}$$

Usually, the elastic modulus of the fibre reinforced concrete, E, is taken as $20\,000\,\text{N/mm}^2$ and Poisson's ratio, ν, is 0.15. At this stage, it is necessary to assume a slab thickness with no real guidance. As a trial, consider 225 mm. Equation (4.5) can now be evaluated as:

$$L = \left(\frac{20\,000 \times 225^3}{12(1 - 0.0225)0.027} \right)^{0.25} = 921\,\text{mm}$$

The distance between the container corner castings is 1500 mm so the ratio S/L is $1500/921 = 1.63$. From Fig. 88, $M_t/P = 0.028$.

Equation (4.6) can now be used to determine the additional stress caused by one corner casting on its neighbour:

$$\sigma_{\text{add}} = \frac{M_t}{P} \frac{6}{h^2} P_2$$

In this case, each container corner casting load is 66 kN, so:

$$\sigma_{\text{add}} = 0.028 \frac{6}{225^2} 66\,000 = 0.22\,\text{N/mm}^2$$

This stress is in fact added twice because there are three corner castings near each other which means one of the corner castings is subjected to the proximity effect twice.

Use Equation (4.7), where σ_{flex} (characteristic flexural strength) is used in place of σ_{char}:

$$\sigma_{max} = \sigma_{flex} - \sigma_{add}$$

$$\sigma_{max} = 2.2 - 2 \times 0.22 = 1.76\,\text{N/mm}^2$$

Now use a form of Equation (4.8) to determine the Single Point Load which in this case accounts for the proximity of two neighbouring corner castings as well as the fatigue effect:

$$\text{SPL} = \text{APL}(\sigma_{flex}/\sigma_{max})$$

$$\text{SPL} = 66(2.20/1.76) = 82.5\,\text{kN}$$

If the effective radius of contact of the corner castings had fallen outside the range 150 mm to 250 mm, Table 4.4 would have been used to adjust the SPL to obtain the Equivalent Single Point Load (ESPL). In this case, it is assumed that the ESPL and the SPL are similar.

Having determined the ESPL, the design chart in Fig. 94 (repeated opposite) can now be used to determine the thickness of the hardstanding. The Design Chart is reproduced here with the 82.5 kN value brought across to the 1 500 000 repetitions curve.

5.2.3 Assessment of vehicle patch loads using modified Westergaard equations

Now consider the highway vehicle and calculate the stresses in the slab when subjected to a patch load of 90 kN. The three Westergaard equations are reproduced here.

(*a*) Patch load within slab:

$$\sigma_{max} = \frac{0.275(1+\nu)}{h^2} P \log\left(\frac{0.36\,Eh^3}{Kb^4}\right) \qquad (4.2)$$

(*b*) Patch load at edge of slab:

$$\sigma_{max} = 0.529(1+0.54\nu)\frac{P}{h^2} \log\left(\frac{0.20\,Eh^3}{Kb^4}\right) \qquad (4.3)$$

(*c*) Patch load at slab corner:

$$\sigma_{max} = \frac{3P}{h^2}\left[1 - \left(\frac{1.41b}{\{Eh^3/[12(1-\nu^2)K]\}^{0.25}}\right)^{0.6}\right] \qquad (4.4)$$

where

σ_{max} = flexural stress (in N/mm^2)
P = patch load (in N), i.e. characteristic wheel load $\times$ load factor $= 90\,000\,\text{N}$

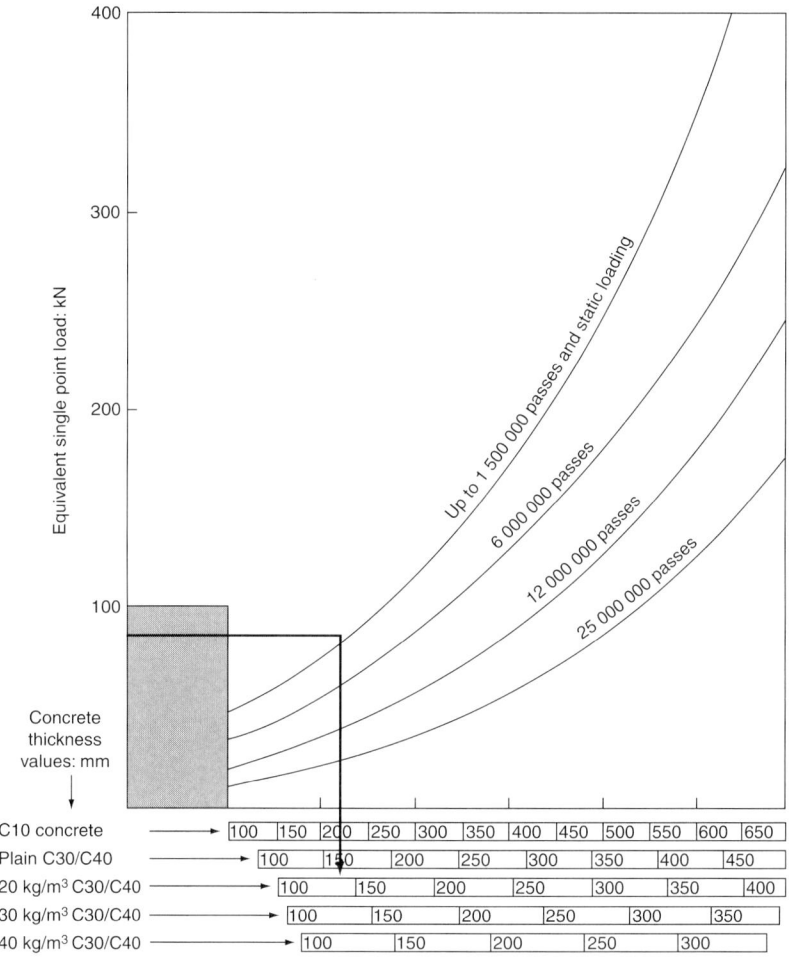

ν = Poisson's ratio, 0.15

h = slab thickness, 225 mm

E = elastic modulus, 20 000 N/mm^2

K = modulus of subgrade reaction, 0.044 N/mm^3

b = radius of tyre contact zone (in mm)

p = contact stress between wheel and hardstanding, 0.8 N/mm^2

= $(W/\pi p)^{1/2} = (90\,000/0.8\pi)^{1/2} = 190$ mm.

The following tables can be used as an aid to evaluating the three Westergaard equations. They apply to the common ν and E values above. Westergaard stresses are obtained by multiplying together two values read from the appropriate tables.

5.2.4   Tables for patch load within slab

Table 5.2.   Values for $[0.275(1 + \nu)/h^2]P$. The underlined figure relates to this example

Load: N	Slab thickness: (in mm)								
	150	175	200	225	250	275	300	325	350
10 000	0.140	0.103	0.079	0.062	0.051	0.042	0.035	0.030	0.026
20 000	0.280	0.206	0.158	0.124	0.102	0.084	0.070	0.060	0.052
30 000	0.420	0.309	0.237	0.186	0.153	0.126	0.105	0.090	0.078
40 000	0.562	0.413	0.316	0.250	0.202	0.167	0.140	0.120	0.103
50 000	0.703	0.516	0.395	0.312	0.253	0.209	0.176	0.150	0.129
60 000	0.840	0.618	0.474	0.375	0.304	0.251	0.211	0.180	0.155
70 000	0.984	0.723	0.553	0.437	0.354	0.293	0.246	0.210	0.181
80 000	1.124	0.826	0.632	0.500	0.404	0.334	0.280	0.240	0.206
90 000	1.265	0.929	0.712	0.562	0.455	0.376	0.316	0.269	0.232
100 000	1.406	1.032	0.790	0.624	0.506	0.418	0.352	0.300	0.258
150 000	2.109	1.548	1.185	0.936	0.759	0.627	0.487	0.450	0.387
200 000	2.812	2.064	1.580	1.248	1.012	0.836	0.704	0.600	0.516

Table 5.3.   Values for $\log(0.36Eh^3/Kb^4)$*. The underlined figure relates to this example*

Load: N	Slab thickness (in mm)								
	150	175	200	225	250	275	300	325	350
10 000	4.544	4.746	4.920	5.073	5.210	5.334	5.447	5.552	5.648
20 000	3.944	4.145	4.319	4.472	4.610	4.734	4.847	4.952	5.048
30 000	3.592	3.793	3.967	4.121	4.258	4.382	4.495	4.600	4.696
40 000	3.341	3.541	3.715	3.869	4.006	4.130	4.244	4.348	4.444
50 000	3.145	3.346	3.520	3.673	3.811	3.935	4.048	4.153	4.249
60 000	2.992	3.193	3.366	3.520	3.657	3.782	3.895	4.000	4.096
70 000	2.831	3.031	3.205	3.359	3.496	3.620	3.733	3.838	3.934
80 000	2.740	2.941	3.115	3.269	3.406	3.530	3.643	3.748	3.844
90 000	2.627	2.825	3.002	3.155	3.292	3.417	3.530	3.634	3.731
100 000	2.538	2.738	2.913	3.714	3.203	3.327	3.441	3.545	3.642
150 000	2.193	2.393	2.567	2.721	2.858	2.982	3.096	3.199	3.296
200 000	1.941	2.142	2.316	2.469	2.607	2.731	2.844	2.948	3.045

5.2.5   Tables for patch load at edge of slab

Table 5.4.   Values for $[0.529(1 + 0.54\nu)/h^2]P$. *The underlined figure relates to this example*

Load: N	Slab thickness (in mm)								
	150	175	200	225	250	275	300	325	350
10 000	0.254	0.187	0.143	0.113	0.091	0.076	0.063	0.054	0.047
20 000	0.508	0.374	0.286	0.226	0.182	0.152	0.126	0.108	0.094
30 000	0.762	0.560	0.429	0.339	0.275	0.227	0.191	0.162	0.140
40 000	1.016	0.748	0.572	0.452	0.364	0.304	0.252	0.216	0.188
50,000	1.270	0.933	0.715	0.565	0.457	0.378	0.317	0.271	0.233
60 000	1.524	1.120	0.858	0.678	0.550	0.454	0.382	0.324	0.280
70 000	1.780	1.307	1.000	0.790	0.640	0.529	0.444	0.379	0.327
80 000	2.212	1.496	1.144	0.904	0.728	0.608	0.504	0.432	0.376
90 000	2.287	1.680	1.287	<u>1.017</u>	0.823	0.681	0.572	0.487	0.420
100 000	2.540	1.866	1.430	1.130	0.914	0.756	0.634	0.542	0.466
150 000	3.810	2.799	2.145	1.695	1.371	1.134	0.951	0.813	0.699
200 000	5.080	3.732	2.860	2.260	1.828	1.512	1.268	1.084	0.932

Table 5.5.   Values for $\log(0.20Eh^3/Kb^4)$. The underlined figure relates to this example

Load N	Slab thickness (in mm)								
	150	175	200	225	250	275	300	325	350
10 000	4.290	4.490	4.664	4.818	4.955	5.079	5.193	5.297	5.393
20 000	3.689	3.890	4.064	4.217	4.355	4.479	4.592	4.697	4.793
30 000	3.337	3.538	3.712	3.865	4.003	4.127	4.240	4.344	4.441
40 000	3.085	3.286	3.460	3.613	3.751	3.875	3.988	4.093	4.189
50 000	2.890	3.091	3.265	3.418	3.555	3.679	3.793	3.897	3.994
60 000	2.737	2.938	3.111	3.265	3.402	3.526	3.640	3.744	3.840
70 000	2.575	2.776	2.950	3.104	3.241	3.365	3.478	4.071	3.679
80 000	2.250	2.686	2.860	3.013	3.151	3.275	3.388	3.492	3.589
90 000	2.372	2.572	2.747	2.900	3.037	3.162	3.275	3.379	3.475
100 000	2.282	2.483	2.658	2.811	2.948	3.072	3.186	3.290	3.387
150 000	1.937	2.138	2.312	2.466	2.603	2.727	2.840	2.944	3.041
200 000	1.686	1.886	2.061	2.214	2.351	2.476	2.589	2.693	2.790

5.2.6   Tables for patch load at corner of slab

Table 5.6.   Values for $1 - \{1.41b/[Eh^3/12(1 - \nu^2)K]^{0.25}\}^{0.6}$. *The underlined figure relates to this example*

Load N	Slab thickness (in mm)								
	150	175	200	225	250	275	300	325	350
10 000	0.682	0.704	0.721	0.735	0.748	0.758	0.768	0.776	0.783
20 000	0.609	0.636	0.657	0.674	0.689	0.703	0.714	0.724	0.733
30 000	0.559	0.588	0.612	0.632	0.649	0.664	0.677	0.688	0.699
40 000	0.519	0.551	0.577	0.599	0.618	0.634	0.648	0.660	0.671
50 000	0.485	0.520	0.548	0.571	0.591	0.608	0.624	0.636	0.648
60 000	0.457	0.494	0.523	0.547	0.568	0.586	0.603	0.617	0.629
70 000	0.426	0.464	0.496	0.522	0.544	0.563	0.580	0.594	0.608
80 000	0.407	0.448	0.480	0.507	0.529	0.549	0.560	0.582	0.596
90 000	0.384	0.426	0.459	<u>0.487</u>	0.511	0.531	0.550	0.565	0.579
100 000	0.365	0.407	0.442	0.471	0.495	0.516	0.536	0.551	0.566
150 000	0.284	0.333	0.371	0.404	0.432	0.455	0.477	0.495	0.511
200 000	0.219	0.272	0.314	0.350	0.380	0.406	0.429	0.449	0.467

Table 5.7.   Values for $3P/h^2$. The underlined figure relates to this example

Load N	Slab thickness (in mm)								
	150	175	200	225	250	275	300	325	350
10 000	1.333	0.980	0.750	0.593	0.480	0.397	0.333	0.284	0.245
20 000	2.667	1.959	1.500	1.185	0.960	0.793	0.667	0.568	0.490
30 000	4.000	2.938	2.250	1.778	1.440	1.190	1.000	0.850	0.735
40 000	5.333	3.918	3.000	2.370	1.920	1.587	1.333	1.136	0.980
50 000	6.667	4.898	3.750	2.963	2.400	1.983	1.667	1.420	1.224
60 000	8.000	5.877	4.500	3.555	2.880	2.380	2.000	1.704	1.469
70 000	9.333	6.857	5.250	4.148	3.360	2.777	2.333	1.988	1.714
80 000	10.667	7.836	6.000	4.740	3.840	3.174	2.667	2.272	1.959
90 000	12.000	8.816	6.750	5.333	4.320	3.570	3.000	2.556	2.204
100 000	13.333	9.796	7.500	5.926	4.800	3.967	3.333	2.840	2.449
150 000	20.000	14.694	11.250	8.889	7.200	5.950	5.000	4.260	3.673
200 000	26.667	19.592	15.000	11.852	9.600	7.934	6.667	5.680	4.898

5.2.7 Solution

The stresses developed in the slab by a patch load of 90 kN can now be calculated using the values underlined in Tables 5.2 to 5.7 as follows.

(a) Maximum stress within the slab $= 0.562 \times 3.155 = 1.77 \, \text{N/mm}^2$.
(b) Maximum stress at slab edge $= 1.017 \times 2.900 = 2.95 \, \text{N/mm}^2$.
(c) Maximum stress at corner of slab $= 0.487 \times 5.333 = 2.60 \, \text{N/mm}^2$.

In the case of edge and corner stresses, aggregate interlock and fibre load transfer may allow the sharing of patch loads by neighbouring slabs. This transfer reduces edge and corner stresses by an amount which depends upon the amount by which the joints open. The true joint opening depends upon many factors, some of which will be unknown at design stage. Experience indicates that significant load transfer occurs when joints open by up to 1 mm. When joint openings exceed 2 mm, there is little or no load transfer. In the absence of other data, Table 5.8 may be used to assess load transfer. The table was developed from the Author's experience in the design and maintenance of hardstandings.

In this example, joints are at 6 m spacings in both directions, so that a maximum 30% load transfer can be assumed. This applies to both edge and corner loading. Therefore, the three critical stresses are as follows:

(a) Maximum stress within the slab $= 1.77 \, \text{N/mm}^2$.
(b) Maximum stress at slab edge $= 2.07 \, \text{N/mm}^2$.
(c) Maximum stress at corner of slab $= 1.82 \, \text{N/mm}^2$.

The characteristic strength of the concrete is $2.2 \, \text{N/mm}^2$ so the proposed slab is satisfactory in the case of patch loading.

5.3 Engineering data for fibre reinforced concrete

The following test data have been obtained as the use of steel and polypropylene fibres has increased.

Table 5.8. Dowel bar load transfer efficiency values

Joint spacing: m	Load transfer: %
6 or less	30
8	20
10	10
12	0

5.3.1   Data relating to steel fibre reinforced concrete

Several research programmes have been undertaken on slabs incorporating steel fibre reinforced concrete. Two such works are summarized below.

(*a*) Imperial College, London (Table 5.9)

- 12 slabs tested (1 m×1 m×50 mm)
- Simply supported on four edges and loaded in the centre
- Same basic concrete mixture
- Slabs contain varying quantity of hooked 60 mm long, 1.0 mm diameter fibres

(*b*) Thames Polytechnic, London (Table 5.10)

Table 5.9.   Summary of results of work undertaken at Imperial College, London

Fibre dosage: kg/m^3	0	20	25	30
First-peak load: kip	2.09	2.14	2.47	2.47
First-peak stress: psi	638	652	754	754
Maximum load: kip	2.11	2.79	2.97	3.10
Plateau load: kip[a]	0	1.96	2.36	2.47

[a] plateau load corresponds to a state of increasing deflection under constant load and represents the collapse load of the slab. 1 kip = 1000 lbf.

Table 5.10.   Summary of results of work undertaken at Thames Polytechnic, London

	Steel fibre type				
	None	Length 60 mm, diam. 1 mm	Length 60 mm diam. 1 mm	Length 60 mm diam. 0.8 mm	Length 60 mm diam. 0.8 mm
Quantity: kg/m^3	0	20	30	20	30
Load at first visible crack: kip	40.5	47.2	53.9	58.4	65.2
Maximum load: kip	45.0	73.1	76.4	87.7	77.6[a]

[a] In the first test the capacity of the testing apparatus was not sufficient; for subsequent tests the apparatus was modified.

- Nine slabs tested ($3\,m \times 3\,m \times 150\,mm$)
- $K = 0.35\,N/mm^3$
- Loaded until failure with point load in centre of slab
- Load at first visible crack also recorded.

5.3.2   Data relating to polypropylene fibre reinforced concrete
The data that follow were obtained at Signet Laboratory and at San Jose State University using plain concrete and concrete with Fibermesh polypropylene fibres added. The concrete had a cement content of $240\,kg/m^3$, water/cement ratio of 0.64 and maximum aggregate size of 20 mm. The following results were obtained.

5.3.2.1   Rate of gain of compressive strength.   Table 5.11 shows the rate of gain of compressive strength for both plain and polypropylene reinforced concrete. Fibre was added at the rate of $890\,g/m^3$.

5.3.2.2   Effect of fibre dosage on compressive strength.   Table 5.12 shows how three levels of fibre dosage influence compressive strength at ages of 7 and 28 days.

Table 5.11.

Age of concrete	Compressive strength: N/mm^2	
	Plain concrete	Polypropylene fibre reinforced concrete
18 h	2.5	2.6
24 h	3.3	3.4
7 days	9.8	10.4
28 days	16.0	17.4

Table 5.12.

Polypropylene fibre dosage: g/m^3	Compressive strength: N/mm^2	
	7 day strength	28 day strength
0	14.2	27.3
600	15.9	29.7
1200	16.2	30.4

Table 5.13.

Polypropylene fibre dosage: g/m^3	Flexural strength: N/mm^2		Tensile strength: N/mm^2	
	7 days	28 days	7 days	28 days
0	1.8	2.5	1.2	1.9
890	2.0	2.7	1.4	2.0

5.3.2.3   Effect of fibre dosage on flexural strength and tensile strength.
Table 5.13 shows flexural strength values determined by bending beams
of dimensions 150 mm×150 mm×450 mm and tensile strength values
obtained by undertaking cylinder splitting tests.

5.4   Long strip hardstanding construction case study

An example of a hardstanding constructed conventionally by the long
strip method is presented in Figs 102 to 118. The hardstanding was
constructed during 1999. A conventionally reinforced 150 mm thick
concrete slab was constructed over a dolomitic limestone 150 mm thick
sub-base. The hardstanding was constructed over a period of three
months. One or two strips running from one side of the building to the
other were constructed each working day. The reinforcement was placed
near the upper surface of the slab and three types of joint were provided.
The joints between neighbouring strips comprised conventional ties
connecting the strips. The transverse joints comprised dowelled joints and
tied joints. In the case of the dowelled joints, sleeves were placed over the
bars at one side of the joint to allow contraction. The ties comprised a
short length of mesh reinforcement placed at the underside of the slab.
Figure 119 illustrates a typical slump test.

*Fig. 102. Typical construction site. Note that the sub-base must support the
weight of the readymix concrete delivery truck*

Fig. 103. This bay is ready for concreting

Fig. 104. The concrete is being compacted whilst the operative ensures that there is a little concrete surcharge in front of the twin beam compactor

Fig. 105. Making good the concrete surface prior to applying the brushed finish

Fig. 106. Finishing the surface with the float prior to applying the brushed finish

Fig. 107. The next bay is ready for concreting as finishing continues on the previous bay

Fig. 108. A small excavator helps in placing the concrete. Because hardstanding concrete is of low workability, this saves much labour

Fig. 109. The excavator allows concrete to be placed at a rate of 15 m³ per hour

Fig. 110. It is common for industrial hardstandings to be extended and care needs to be exercised to balance the needs of the contractor with those of the hardstanding operator

Fig. 111. A poker vibrator ensures that compaction is achieved to full depth. Care must be undertaken to avoid over vibrating which will lead to segregation

Fig. 112. A brushed finish is achieved by this combination float/brush device

Fig. 113. The twin beam compactor is winched along progressively. The winch is attached to a heavy item of plant at each side of the twin beam compactor

Fig. 114. The motor powers a rotating shaft which has eccentric weights positioned regularly to ensure compaction takes place uniformly across the bay

Fig. 115. The winching of the twin beam compactor is coordinated with the maintaining of surcharge to ensure uniform compaction

Fig. 116. Before the readymix truck returns to the public road, the concrete adhering to its drum and discharge chute is washed away using the truck's on-board water supply. It is important to undertake this in an environmentally acceptable manner. The residue must not be washed into the drainage system. Any surplus concrete must be disposed of carefully

Fig. 117. It is common for safety fencing to be fixed to the surface of the concrete using expanding bolts

Fig. 118. It is inevitable that trenches will need to be formed in the hardstanding during its life. To attempt to minimize disruption later, ducts should be included where possible to permit future installation of services with as little disturbance as possible to the surface

Fig. 119. Typical slump test result for pavement-quality concrete

6. Specification of industrial hardstandings

6.1 Introduction

This chapter presents a set of model contract documents which can be used as a guide for a typical in-situ industrial hardstanding construction. It is assumed that a hardstanding needs to be upgraded so the existing concrete needs to be removed and a new reinforced concrete slab is to be installed. The documents comprise the following.

- Instructions for tendering
- Brief description of works
- Conditions of contract
- Specification
- Preamble and bill of quantities
- Form of Quotation

Anyone wishing to receive the documentation as a Microsoft Word file should contact the Author by e-mail (John.Knapton@ncl.ac.uk)

6.2 Instructions for tendering

TENDERS MUST BE SUBMITTED IN ACCORDANCE WITH THE FOLLOWING INSTRUCTIONS. TENDERS NOT COMPLYING WITH THESE INSTRUCTIONS MAY BE REJECTED BY THE EMPLOYER WHOSE DECISION IN THE MATTER SHALL BE FINAL.

1. The tender shall be made on the Form of Quotation incorporated in the tender documents. It shall be signed and submitted with the Bill of Quantities which shall be fully priced and totalled in ink. The quote shall be based on the Specification, Drawings and the Bill of Quantities.
2. The tender documentation shall be treated as private and confidential.

3. Tenders shall not be qualified and shall be submitted strictly in accordance with the tender documents. Only unqualified tenders shall be considered for acceptance. However, any alterations or qualifications which the tenderer wishes to be considered should be made in an accompanying letter with any financial consequences.

4. Tenders shall be submitted exclusive of VAT.

5. Tenderers are expected to visit the site and appraise the content of the work. Arrangements to visit site should be made through Mr. B. Field who can be contacted by telephone on: 0191 222 6407.

6. Tenders shall be forwarded to the offices of John Knapton Consulting Engineers, City Road, Newcastle upon Tyne, no later than noon on 2nd September 1999.

7. The Employer does not bind himself to accept the lowest or any tender.

8. The tenderer should note that the works will not commence on site until January 2000 and will be completed by the end of March 2000. The contract is a fixed price quotation and does not include any price fluctuation clause.

6.3   Brief description of works

THIS INFORMATION IS FOR TENDERING PURPOSES ONLY AND DOES NOT CONSTITUTE PART OF ANY CONTRACT.

The works comprise the following.

1. The excavation and removal of the existing concrete slab and its removal from site.

2. The replacement of the in-situ concrete slabs including all associated transverse, induced, longitudinal and expansion joints and reinforcing materials.

3. Any damaged sub-base to the excavated slabs to be removed and replaced with DTp Specification for Highway Works Clause 803 Type 1 sub-base material and compacted by vibratory roller as specified in Clause 802 of the same document.

6.4   Conditions of Contract

This project shall be carried out according to the Institution of Civil Engineers' Conditions of Contract for Minor Works, 2nd Edition (March 1998) as approved by the Institution of Civil Engineers and The Association of Consulting Engineers.

6.4.1 Institution of Civil Engineers Conditions of Contract for Minor Works

6.4.2 Appendix to the Conditions of Contract
This is to be written by each prospective tenderer and shall comprise the following:

1. Short description of the work to be carried out under the Contract under the heading:

    Replacement and Repair to Concrete Slabs.

2. Payment to be made under Article 2 of the Agreement in accordance with Clause 7 will be calculated on the following basis.
3. Where a Bill of Quantities or a Schedule of Rates is provided the method of measurement used is: (insert as required).

PREAMBLE

4. Name of the Engineer (Clause 2.1)
   JOHN KNAPTON
5. Starting date (if known) (Clause 4.2)
   2nd January 2000
6. Period for completion (Clause 4.2)
   12 weeks
7. Liquidated damages (Clause 4.6)
   £600/day
8. Limit of liquidated damages (Clause 4.6)
   £15 000
9. Defects Correction Period (Clause 5.1)
   12 months
10. Rate of retention (Clause 7.3)
    5%
11. Limit of retention (Clause 7.3)
    £15 000
12. Minimum amount of interim certificate (Clause 7.3)
    £20 000
13. Bank whose base lending rate is to be used (Clause 7.8)
    Byker Peoples Credit Union
14. Insurance of the Works (Clause 10.1)
    Required — enter name of insurer

15.   Minimum amount of third party insurance (persons and property) (Clause 10.6)
£5M
Any one accident/number of accidents unlimited
16.   Name and address of the Planning Supervisor (Clause 13(1)(b))
Bernard Field
19 City Road
Newcastle upon Tyne NE1
17.   Name of the Principal Contractor (Clause 13(1)(b))
Company's name and address to be inserted here.

6.4.3   ICE Conditions of Contract for Minor Works
CONTRACT SCHEDULE
(List of documents forming part of the Contract)

- The Agreement
- The Contractor's Tender Price (excluding any general or printed terms contained or referred to therein unless expressly agreed in writing to be incorporated in the Contract)
- The Conditions of Contract
- The Appendix to the Conditions of Contract
- The Drawings. Reference numbers
- As listed on page XX of the tender document
- The Specification
- The priced Bill of Quantities
- The Schedule of Rates
- The Daywork Schedules

6.5   Specification
The specification should include details on the following:

- General
- Traffic safety
- Cold weather workings
- Saw cutting
- Sub-base
- Concrete
- Reinforcement
- Curing
- Tolerances of concrete slab
- Joint sealants and filler

These points are now considered in turn.

6.5.1   Specification General
Noise
1. The Contractor shall comply with the general requirements set out in BS 5228: Parts 1 and 2: 1984 *Noise control on construction and open sites*.

Setting Out
2. The Contractor's method of establishing setting out points must ensure that they are maintained and that replacement concrete bays comply with the agreed dimensions on the drawings.

Dust and mud on highways
3. The Contractor shall take all reasonable steps to minimize dust and/or mud nuisance during the construction of the works.
4. All public highways used by vehicles of the Contractor or any of his Sub-contractors or suppliers of materials or plant shall be kept clean and clear of all dust and mud dropped from those vehicles or their tyres. Similarly, all dust and mud from the works spreading on public highways shall be immediately cleared at the expense of the Contractor.
5. Clearance shall be effected immediately by manual sweeping and removal of debris, or, if so directed by the Engineer, by mechanical sweeping and cleaning equipment and all dust, mud and other debris shall be removed entirely from the road surface. Additionally, if so directed by the Engineer, the road surface shall be hosed or watered using suitable equipment.
6. Compliance with the foregoing will not relieve the Contractor of any responsibility for complying with the requirements of the Highway Authority in respect of keeping roads clean.

Programme of works
In accordance with the Conditions of Contract, the Contractor shall submit a firm programme of works as soon as practicable after the acceptance of his tender. In addition, the Contractor shall provide all subsequent revisions which may be required by the Engineer. Consultations with the Supervising Officer and the employer will be necessary to provide 7 days notice of the Contractor's intention to enter any of the employer's premises to enable the operations of the employer to be modified to accommodate the works programme.

Permits
7. The Supervising Officer will issue a permit to the Contractor's Agent who, after establishing his own identity, shall vouch for all

employees of the Contractor under his charge of control. On commencement of the works, the Contractor's Agent shall ensure that he is in possession of the necessary permit at all times.

8. Security checks may be made from time to time and the Contractor's employees will be required to show proof of identity and of their need to be on the Employer's property to any responsible member of the Employer's staff.

9. The Contractor's Agent shall ensure the return of all permits when the need to be on the site ceases.

Health and safety

10. The Contractor, when carrying out the works, shall comply with all requirements of The Health & Safety at Work Act 1994, so far as they affect personnel who are required to undertake the work on site.

   The Contractor's attention is also drawn to the Construction (Design and Management) (CDM) Regulations 1994, and the Contractor shall provide all details required under these Regulations such as Health & Safety Policy, Risk Assessment and Training Certificates of personnel, etc. All requirements shall be provided prior to the works commencing and the Contractor shall comply with all requests of the Planning Supervisor.

List of drawings

11. The drawings which form part of the Contract are as follows:

   99/100/1   Existing
   99/100/2   Proposed
   99/100/3   Location Plan.

6.5.2   Traffic Safety

1. The Contractor shall be responsible for the erection, maintenance and repositioning of appropriate traffic signs to ensure the safety of all vehicles, pedestrians and maintenance associated with the Employer's operations. Where appropriate, all signing shall conform to the requirements of Chapter 8 of the Traffic Signs Manual published by HMSO. The Contractor shall ensure that all employees are familiar with Chapter 8 and shall observe the rules and regulations enforced within the Employer's Safety Book.

2. All traffic control proposals shall be approved by the Employer's Supervising Officer prior to commencement of the works.

3. All excavations shall be effectively protected from all road users, pedestrians and operations implemented by the client. The Contractor shall be responsible for maintenance of all guards and barriers during the contract period.
4. All personnel of the Contractor or his Sub-contractor must at all times wear approved reflective or fluorescent clothing.
5. All precautions shall be taken to ensure that all roads and hardstandings, both within and outside of the site, are kept clean of any spillage or debris occurring from the works. Any such spillage or debris shall be cleaned immediately.

6.5.3   Cold weather working
1. No hardstanding construction materials shall be incorporated in the Works in a frozen condition.
2. Materials for use in the hardstanding shall not be laid on any surface which is frozen or covered with ice.
3. The temperature of concrete or cement-bound material in any pavement layer shall be not less than 5°C at the point of delivery. These materials shall not be laid when the air temperature in the shade falls below 3°C and laying shall not be resumed until the rising air temperature in the shade reaches 5°C.
4. If frost occurs during the first 20 days after placing the concrete slabs, one day shall be added to the period which would otherwise be required before opening to traffic for each night on which the temperature of the surface of the layer in question falls to 0°C or below.
5. For the use of hot-applied sealants, the temperature of the pavement receiving the sealant shall be not less than 5°C.

6.5.4   Saw Cutting
1. The excavation shall commence by saw cutting vertically around the perimeter of the excavation to the full depth of the concrete to be removed.
2. All excavated materials shall be removed, the bottom of the excavation swept of loose material and all dust arising from the saw cutting operation shall be blown out using compressed air.

6.5.5   Sub-base
Granular Sub-base Material DTp Type 1
1. DTp Type 1 granular sub-base material shall be crushed rock, crushed slag, crushed concrete or well burnt non-plastic shale. The

Table 6.1.   Grading limits for DTp Type 1 Sub-base Material. Particle size shall be determined by the washing and sieving method of BS 812 Part 103

BS sieve size	Percentage by mass passing
75 mm	100
37.5 mm	85–100
10 mm	40–70
5 mm	25–45
600 μm	8–22
75 μm	0–10

material shall be well graded, and lie within the grading envelope of Table 6.1.

2. The material passing the 425 μm BS sieve shall be non-plastic as defined by BS 1377: 1990, and tested in compliance therewith.
3. The material shall be transported, placed and compacted without drying out or segregating.
4. The material shall have a 10% fines value of 50 KN or more when tested in accordance with BS 812, except that samples shall be tested in a saturated and surface dried condition. Prior to testing, the selected test items shall be soaked in water at room temperature for 24 hours without previously having been oven dried.

Compaction

1. Compaction shall be completed as soon as possible after the material has been spread and in accordance with the requirements for individual materials.
2. Special care shall be taken to obtain full compaction in the vicinity of both longitudinal and transverse joints.
3. Compaction of unbound materials shall be carried out by a method specified in Table 6.2, unless the Contractor demonstrates at site trials that a state of compaction achieved by an alternative method is equivalent to or better than that using the specified method. The procedure for these trials shall be subject to approval by the Supervising Officer.
4. The surface of any layer of material shall, on completion of compaction and immediately before overlaying, be well closed,

Table 6.2.   Compaction requirements for DTp granular material Type 1

Type of compaction plant	Category	Number of passes for layers not exceeding the following compacted thicknesses (in mm)		
		110	150	225
Smooth-wheeled roller (or vibratory roller operating without vibration	Mass per metre width or roll: over 2700 kg up to 5400 kg over 5400 kg	16 8	Unsuitable 16	Unsuitable Unsuitable
Pneumatic-tyred roller	Mass per wheel: over 4000 kg up to 6000 kg over 6000 kg up to 8000 kg over 8000 kg up to 12 000 kg over 12 000 kg	12 12 10 8	Unsuitable Unsuitable 16 12	Unsuitable Unsuitable Unsuitable Unsuitable
Vibratory roller	Mass per metre width of vibrating roll: over 700 kg up to 1300 kg over 1300 kg up to 1800 kg over 1800 kg up to 2300 kg over 2300 kg up to 2900 kg over 2900 kg up to 3600 kg over 3600 kg up to 4300 kg over 4300 kg up to 5000 kg over 5000 kg	16 6 4 3 3 2 2 2	Unsuitable 16 6 5 5 4 4 3	Unsuitable Unsuitable 10 9 8 7 6 5
Vibrating plate compactor	Mass per square metre of base plate: over 1400 kg/m^2–1800 kg/m^2 over 1800 kg/m^2–2100 kg/m^2 over 2100 kg/m^2	8 5 3	Unsuitable 8 6	Unsuitable Unsuitable 10
Vibro-tamper	Mass: over 50 kg up to 65 kg over 65 kg up to 75 kg over 75kg	4 3 2	8 6 4	Unsuitable 10 8
Power rammer	Mass: 100 kg up to 500 kg over 500 kg	5 5	8 8	Unsuitable 12

free from movement under compaction plant and from ridges, cracks, loose material, pot holes, ruts or other defects. All loose, segregated or otherwise defective areas shall be removed to the full thickness of the layer and new material laid and compacted.

5. For the purpose of Table 6.2, vibrating-plate compactors are machines having a base plate to which is attached a source of vibration consisting of one or two eccentrically weighted shafts.

(a) The mass per square metre of base plate of a vibrating plate compactor is calculated by dividing the total mass of the machine in its working condition by its area in contact with the material to be compacted.

(b) Vibrating-plate compactors shall be operated at the frequency of vibration recommended by the manufacturer. They shall normally be operated at travelling speeds of less than 1 km/h but if higher speeds are necessary, the number of passes shall be increased in proportion to the increase in speed of travel.

6.5.6 Concrete
Pavement-quality concrete

1. All concrete in the slabs shall comprise air-entrained C30 grade concrete to BS 5328, with a minimum of 320 kg/m^3 OPC content.

2. The water content shall be the minimum required to provide the required workability for full compaction of the concrete as determined by trial mixing or approved means, and the maximum free water/cement ratio shall be 0.5.

3. Aggregates for all pavement-quality concrete shall be naturally occurring materials complying with BS 882.
    The nominal coarse aggregate size shall not exceed 40 mm.
    The chloride ion content of the aggregate to be used in concrete with embedded metal, determined in accordance with BS 812 shall satisfy the requirements given in Appendix C of BS 882.
    Fine aggregate containing more than 25% by mass of acid soluble material, as determined in BS 812, in either the fraction retained on or the fraction passing the 600 μm BS sieve shall not be used in the top 50 mm of slabs.

4. An air-entraining admixture complying with BS 5075, Part 2 shall be used in the concrete in at least the top 50 mm of slabs. Admixtures containing chloride shall not be used. The total quantity of air in air-entrained concrete as a percentage of volume of the mix shall be as follows, within the tolerances given in BS 5328.

Nominal aggregate size: mm	Air content: %
20	4.5 to 5.5
40	4 to 6

The air content shall be determined at the site by a pressure type air meter in accordance with BS 1881, at the rate of one determination to each 20 m length of slab or less constructed at any one time or at least three times per day. If the air content is outside the specified limits a further test shall be made immediately on the next available load of concrete before discharging. If the air content is still outside the limit, the contractor shall immediately adjust the air content of the concrete or improve its uniformity before any further concrete is used in the works.

Air-entraining agent shall be added at the batcher by an apparatus capable of dispensing the correct dose within the limits given in BS 5328 and so as to ensure uniform distribution of the agent throughout the batch during mixing.

5. The density of concrete grade C30 shall be such that the total air voids shall not be more than 8% for 20 mm aggregate or 7% for 40 mm aggregate.

6. Sampling and testing and compliance for the specified characteristic strength of concrete mixes during the works shall be in accordance with BS 5328, except that it shall be at the following rate of sampling and testing and with the following requirements.

Concrete shall be supplied from a QSRMC Registered Plant. The Contractor shall supply details of the Ready Mix Concrete Supplier and Mix Design for approval and shall allow for trial mixes to be undertaken and testing to be carried out to determine the suitability of the concrete.

Three number 150 mm cubes shall be made and cured and tested in accordance with BS 1881 from concrete delivered onto the site. At least one batch of cubes shall be made for 150 m^3 of concrete slab or for each day concrete is to be poured.

One cube shall be tested at 7 days and one cube at 28 days by a reputable company and the results shall be presented to the Supervising Officer for assessment.

The remaining cube shall be retained until all results are available and be crushed if instructed to provide a comparison of strength.

7. The workability of the concrete at the point of placing shall enable the concrete to be fully compacted and finished without undue

flow. The workability shall be determined by the slump test in accordance with BS 1881 for each concrete load delivered to site and shall be maintained at the optimum level within a tolerance in accordance with BS 5328.

Placing and compaction

1. The concrete shall be spread uniformly without segregation or varying degrees of pre-compaction, by conveyor, chute or by other means approved by the Supervising Officer. The concrete shall be struck off by a screed so that the average and differential surcharge is sufficient to ensure that after compaction the surface is to the required levels. The concrete shall be compacted by vibrating finishing beams. In addition, internal poker vibration shall be used for slabs thicker than 200 mm and may be used for lesser thicknesses. When used, the pokers shall be at points not more than 500 mm apart over the whole area of the slab, and adjacent to the side forms or the edge of a previously constructed slab.

    The surface shall be regulated and finished to the top of the forms or adjacent slab or pavement layer by using twin vibrating finishing beams. The beams shall be metal with a contact face at least 50 mm wide and a vibrating unit having a minimum centrifugal force of 4 kN with the frequency recommended by the manufacturer or an equivalent compactive effort. The vibrating beams shall be moved forward at a steady speed of 0.5 m to 1 m per minute whilst vibrating over the compacted surface to produce a smooth finish.

    Joint grooves shall be constructed in compliance with Specification Clause 6.5.3. Any irregularities at wet formed joint grooves shall be rectified by means of a vibrating float at least 1.0 m wide drawn along the line of the joint. The whole area of the slab shall be regulated by two passes of a scraping straight edge not less than 1.8 m wide or by a further application of a twin vibrating finishing beam. Any excess concrete on top of the groove former shall be removed before the surface is textured.

    The concrete shall be placed and compacted within the time to completion given in Table 6.3.

    After the final regulation of the surface of the slab and before the application of the curing membrane, the surface of concrete slabs shall be brush textured in a direction at right angles to the longitudinal axis of the bay. The brushed surface texture shall be applied evenly across the slab in one direction by the use of a wire brush not less than 450 mm wide. The brush shall be made of 32 gauge tape wires grouped together in tufts spaced at 10 mm

Table 6.3. Maximum working times (in hours)

Temperature of concrete at discharge	Reinforced concrete slabs constructed in two layers slabs without retarding		All other concrete slabs	
	Mixing to finishing concrete	Between layers	Mixing to finishing concrete	Between layers in two-layer work
Not more than 25°C	2	$\frac{1}{2}$	3	$1\frac{1}{2}$
Exceeding 25°C but not exceeding 30°C	2	$\frac{1}{2}$	2	1
Exceeding 30°C	Unacceptable for paving	—	Unacceptable for paving	—

centres. The tufts shall contain an average of 14 wires and initially be 100 mm long. The brush shall have two rows of tufts. The rows shall be 20 mm apart and the tufts in one row shall be opposite the centre of the gap between the tufts in the other row. The brush shall be replaced when the shortest tufts wears down to 90 mm long. The minimum texture depth shall be an average of 0.75 mm with no measurement less than 0.65 mm.

Joint Grooves
General
1. Joint grooves shall be wet formed or sawn in the surface slabs to promote cracks at the required positions. They may be of any convenient width but their depth shall be as given in this Specification.
2. Transverse joint grooves which are initially constructed to less than the full width of the slab shall be completed by sawing through to the edge of the slab and across longitudinal joints as soon as any forms have been removed and before an induced crack develops at the joint.

Sawn transverse joint groove
3. Sawing shall be undertaken as soon as possible after the concrete has hardened sufficiently to enable a sharp edged groove to be produced without disrupting the concrete and before random cracks develop in the slab. The grooves shall be of depth 50 mm and of width 6 mm.

Wet formed transverse joint grooves
4. Grooves shall be formed in the plastic concrete prior to the final regulation and finishing of the surface, either by vibrating a metal blade into the concrete to the required depth and inserting a proprietary groove former into the groove or the groove former may be vibrated vertically into the plastic concrete.
5. The disturbed concrete shall be fully recompacted around the former on each side and the surface regulated. If grooves to be formed are wider than 15 mm some of the disturbed concrete shall be removed unless it can be shown that surface regularity can be achieved across each joint. The groove former shall then remain in the correct position, alignment and depth below the surface, until temporary or permanent sealing is carried out.

Longitudinal construction joint grooves in surface slabs
6. The grooves shall be formed by fixing a groove forming strip along the upper edge of the slab already constructed before concreting the adjacent slab. Where the edge of the concrete is damaged it shall be ground or made good to the satisfaction of the Supervising Officer before fixing the groove forming strip. Alternatively, the subsequent slab may be placed adjacent to the first and a sealing groove sawn later in the hardened concrete to a depth of 25 mm.

Slip membrane
1. Slip membranes shall be impermeable plastic sheeting 125 μm thick laid flat without creases. Where an overlap of plastic sheets is necessary this shall be at least 300 mm. There shall be no standing water on or under the membrane when the concrete is placed upon it.
2. When necessary, the surface of the sub-base should be lightly blinded with fine material prior to placing the slip membrane.

6.5.7   Steel reinforcement
Fabricated reinforcement
1. Reinforcement shall comply with any of the following standards and be prefabricated sheets, or bars assembled on site and shall be free from oil, dirt, loose rust or scale:
(a) Steel fabric in flat sheets		BS 4483
(b) Hot rolled steel bars grade 250		BS 4449
(c) Hot rolled steel bars grade 460		BS 4449
(d) Cold worked steel bars		BS 4461

2. When deformed bars are used they shall conform to Type 2 bond classification of BS 4449 or BS 4461.
3. Laps in the longitudinal bars shall be not less than 35 bar diameters or 450 mm, whichever is the greater. At laps between prefabricated sheets the first transverse bar of one sheet shall lie within the last complete mesh of the previous sheet. There shall be a minimum of 1.2 m longitudinally between groups of transverse laps or laps in prefabricated reinforcement sheets.
4. Laps in any transverse reinforcement shall be a minimum of 300 mm. Where prefabricated reinforcement sheets are used and longitudinal and transverse laps would coincide, no lap is required in the transverse bars within the lap of longitudinal reinforcement. These transverse bars may be cropped or fabricated shorter so that the requirements for cover are met.
5. The reinforcement is to be positioned prior to concreting and shall be fixed on approved metal supports and retained in position at the required depth below the finished surface and distance from the edge of the slab so as to ensure that the required cover is achieved. Reinforcement assembled on site shall be tied, or firmly fixed, by a procedure agreed with the Supervising Officer, at sufficient intersections to provide the above rigidity.
6. The reinforcement shall be so placed that after compaction of the concrete the cover below the finished surface is 60 ± 10 mm.

Dowel bars

1. Dowel bars shall be Grade 250 steel complying with BS 4449 and shall be free from oil, dirt, loose rust or scale. They shall be straight, free of burrs or other irregularities and the sliding ends sawn or cropped cleanly with no protrusions outside the normal diameter of the bar. Dowel bars shall be 20 mm diameter at 300 mm spacing, 400 mm long for slabs up to 225 mm thick and 25 mm diameter for slabs 225 mm thick or more.
2. Dowel bars shall be supported on cradles in prefabricated joint assemblies positioned prior to construction of the slab.
3. Dowel bars shall be positioned at mid-depth from the surface level of the slab, ±20 mm. They shall be aligned parallel to the finished surface of the centre line of the slab and to each other within the following tolerances.

    (*a*) For bars supported by cradles prior to construction of the slab.

      (i) All bars in a joint shall be within ±3 mm.
      (ii) Two thirds of the bars shall be within ±2 mm.
      (iii) No bar shall differ in the alignment from an adjoining bar

by more than 3 mm in either the horizontal or vertical plane.

(*b*) For all bars after construction of the slab.

(i) Twice tolerances for alignment as above.
(ii) Equally positioned about the intended line of the joint with tolerances of ±25 mm

4. Cradles supporting dowel bars shall not extend across the line of the joint.

5. The assembly of dowel bars and supporting cradles, including the joint filler board in the case of expansion joints shall have the following degree of rigidity when fixed in position.

(*a*) For expansion joints, deflection of the top edge of the filler board shall be no greater than 30 mm, when a load of 1.3 kN is applied perpendicularly to the vertical face of the joint filler board and distributed over a length of 600 mm by means of a bar or timber packing, at mid-depth and mid-way between individual fixings, or 300 mm from either end of any length of filler board, if a continuous fixing is used. The residual deflection after removal of the load shall not be more than 3 mm.

(*b*) The joint assembly fixings to the sub-base shall not fail under a 1.3 kN load applied for testing the ridgity of the assembly but shall fail before the load reaches 2.6 kN.

(*c*) The fixings for the contraction joints shall not fail under 1.3 kN load and shall fail before the load reaches 2.6 kN when applied over a length of 600 mm by means of bar or timber packing placed as near to level of the line of fixings as practicable.

(*d*) Failure of the fixings shall be deemed to be when there is displacement of the assemblies by more than 3 mm with any form of fixing, under the test load. The displacement shall be measured at the nearest part of the assembly to the centre of the bar or timber packing.

6. Dowel bars shall be covered by a thin plastic sheath for at least two thirds of the length from one end for dowel bars in contraction joints or half the length plus 50 mm for expansion joints. The sheath shall be tough, durable and of an average thickness not greater than 1.25 mm. The sheathed bar shall comply with the following pull out test.

Four bars shall be taken at random from stock and without any special preparation shall be covered by sheaths as required in this clause. The halves of the dowel bars which have been sheathed shall be cast centrally into concrete specimens 150 mm × 150 mm × 450 mm, made of the same mix proportions to be used in the pavement, but with the maximum nominal aggregate size of 20 mm and cured in accordance with BS 1881. At 7 days a tensile load shall be applied to achieve a movement of the bar of at least 0.25 mm.

7. For expansion joints, a closely fitted cap 100 mm long consisting of waterproofed cardboard or an approved synthetic material shall be placed over the sheathed end of each dowel bar. An expansion space equal in length to the thickness of the joint filler board shall be formed between the end of the cap and the end of the dowel bar.

6.5.8 Curing

1. Immediately after the surface treatment is complete, the surface and exposed edges of the slabs shall be cured by the application of an approved resin-based aluminized reflecting curing compound.
2. Resin-based aluminized curing compound shall contain sufficient flake aluminium in finely divided dispersion to produce a complete coverage of the sprayed surface with a metallic finish. The compound shall become stable and impervious to evaporation of water from under the concrete surface within 60 m of application and shall have an efficiency index of 90%.
3. The curing compound shall not react chemically with the concrete to be cured and shall not crack, peel or disintegrate within three weeks after application.
4. Prior to application, the contents of any containers shall be thoroughly agitated. The curing compound shall be mechanically applied using a fine spray onto the surface at a rate of 10% above the recommended coverage rate.
5. The mechanical sprayer shall incorporate an efficient mechanical device for continuous agitation and mixing of the compound in its container during spraying.

6.5.9 Tolerances of concrete slabs

1. The longitudinal regularity of the surface of the slab shall be laid to the stated levels to a tolerance of ±6 mm.
2. The surface shall also be placed to ensure that the tolerance of ±3 mm to the stated levels is achieved under a 3 m long straight edge.
3. At the edge of the repair and existing concrete pavement, the maximum allowable tolerance shall be 3 mm between the

pavement surface and the underside of a 3 m long straight edge.

4. If any tolerances are exceeded the Contractor shall provide a Method Statement for the approval of the Supervising Officer to remedy the situation.

6.5.10   Joint sealants and filler

Hot applied sealants

1. Hot-applied sealants shall comply with ASTM D3406, or D3569 (for fuel resistance).

2. Hot-applied sealants complying with BS 2499 shall be used only in a joint between concrete surface slabs and bituminous surfacing.

Cold-applied sealants

3. Cold-applied sealants shall be polysulphide-based sealants complying with BS 5212. The sealant shall consist of a polymer resin and a curing agent. In addition, a separate retarder may be used in accordance with the sealant manufacturer's instructions.

4. The seal shall be durable, elastometric material of low modulus without plasticity after curing for the manufacturer's recommended period. In addition to the manufacturer's Certificate of Compliance with BS 5212, a certificate shall be provided confirming that the 5 s reading on the Short Hardness Scale A, as measured by a meter in accordance with BS 2719, is less than 20° for a cured sample 7 days after mixing. The difference between the Short Hardness measurement at 7 days and the measurement carried out, shall be not more than 5°C.

5. For joints in kerbs and joints other than pavements, gunning grades of two-part polysulphide-based sealants complying with BS 4254 may be used. Alternatively polyurethane-based sealing compounds may be used provided their performance is not inferior to BS 4254 material.

Preformed Compression Seals

6. Preformed compression seals made of polychloroprene elastomers complying with BS 2752 shall conform with the requirements of ASTM D2628. Seals of butadiene–acrylonitrile or other synthetic rubbers may be used if certificates are produced to show that they conform to the performance requirements of ASTM D2628 for oven ageing, oil and ozone resistance, low temperature stiffening and recovery.

7. Seals made of ethylene vinyl acetate in microcellular form and other synthetic materials may be used in longitudinal joints and in other structures if test certificates are produced to show the adequate resistance against ageing when testing in accordance with BS 4443: Part 4, Method 10 and Method 12 respectively. The compression set of any seal shall not be greater than 15% when the specimen is subjected to a 25% compression in accordance with BS 4443: Part 1, Method 6. When immersed in standard oils for 48 h at 25°C in accordance with BS 903: Part A16, the volume change shall not be greater than 5%.

8. Compression seals shall be shaped so that they will remain compressed at all times and shall have a minimum of 20 mm contact face with the sides of the sealing groove. If lubricant–adhesive is used, it shall be compatible with the seal and the concrete and shall be resistant to abrasion, oxidation, fuels and salt.

Dimensions of applied joint seals

9. Joint seals shall be constructed in accordance with Table 6.4 for hot- and cold-applied sealants, a compressible caulking material, debonding strip or tape of a suitable size to fill the width of the sealing groove shall be firmly packed or stuck in the bottom of the sealing groove to provide the correct depth of seal as described in

Table 6.4.   Dimensions of Applied Joint Seals

Spacing of contraction joints: m	Minimum width: mm	Minimum depth of seal: mm		Depth of top of seal below the concrete surface: mm
		Cold-applied	Hot-applied	
15 and under	13	13	15	5±2
Over 15 to 20	20	15	20	5±2
Over 20 to 25	25	20	25	5±2
Over 25	30	20	25	7±2
Expansion all	30	20	25	7±2
Gully/manhole slabs	20	15	20	0±3

Note. The depth of seal is that part in contact with the vertical face of the joint groove. The depth of seal below the surface shall be taken at the centre of an applied seal relative to a short straight edge, 150 mm long, placed centrally across the joint

Table 6.4, with the top of the seal at the correct depth below the surface of the concrete.

Joint filler board

10. Joint filler board for expansion joints and manhole and gully slab joints shall have a thickness of 25 mm within a tolerance of ±1.5 mm and be of firm compressible material or a bonded combination of compressible and rigid materials of sufficient rigidity to resist deformation during the passage of the concrete paving plant. The depth of the joint filler board for manhole and gully slabs shall be the full depth of the slab less the depth of the sealing groove. In expansion joints, the filler board shall have a ridged top. Holes for dowel bars shall be accurately bored or punched out to form a sliding fit for the sheathed dowel bars.

6.6   Preamble and Bill of Quantities

PREAMBLE TO THE QUOTATION

THE TENDER SHALL BE MADE ON THE FORM OF QUOTATION ENCLOSED. IT SHALL BE SIGNED AND SUBMITTED WITH THE SCHEDULE OF WORKS, WHICH SHALL BE FULLY PRICED AND TOTALLED IN INK AND SHALL INCLUDE FOR THE FOLLOWING UNLESS STATED OTHERWISE.

1. Labour and cost in connection therewith.
2. Supply of materials, goods, storage and costs in connection therewith, including delivery to site. Taking delivery of materials and goods supplied by others, unloading, storage and costs in connection therewith.
3. Disposal of unsuitable materials off site. Stockpiling and double handling of materials as necessary.
4. Plant and costs in connection therewith.
5. Fixing, erecting and installing or placing of materials in position, including forming of all joints.
6. Temporary works, including protective fencing and coning of operations.
7. The effect of phasing the works for alterations or additions to accommodate the operations of the Client to the extent that such work is set forth or recently implied in the documents on which the tender is based.
8. General obligations, liabilities and risks involving the execution of works set forth or reasonably implied in the documents on which the tender is based.
9. Waste.

10. Supply and provision of results for tests on materials and workmanship as specified in the tender documentation.
11. Compliance with Quality Assurance standards.
12. Preparation and supply of detailed drawings showing position of all slabs and levels on completion of the work.
13. Compliance with all Health & Safety Requirements of the 1994 CDM Regulations.
14. Provision of automatic level, tripod and staff, and 3 m straight edge and wedge, together with appropriate tapes for temporary use by the Supervising Engineer during the course of the works.

Tables 6.5 to 6.7 set out items to be included in the Bill of Quantities.

Table 6.5.

No.	Description	Unit	Qty	Rate	£
	Preliminaries				
1.	Insurance of Works	Item			
2.	Implementation and maintenance of all phasing and protection of the works	Item			
3.	Traffic control	Item			
4.	Provision of all welfare facilities	Item			
	TOTAL TO QUOTATION				

Table 6.6.

No.	Description	Unit	Qty	Rate	£
	Reference is made to drawing Nos 99/100/1 and 99/100/2				
1.	Excavate and dispose off site existing concrete slabs and sand bed	m^3	1400		
2.	Full depth saw cut through existing concrete slab approximate depth 225 mm and clean vertical face.	m	62		
3.	Full depth saw cut existing salvacim pavement approximate depth 225 mm and clean vertical face.	m	62		
4.	Reinstate existing sub-base and place and compact regulating Type 1 sub-base to level	m^2	4242		
5.	Provide and place 225 mm thick pavement concrete on separation membrane including all joints as shown on the drawings and in accordance with the Specification	m^2	4242		
	TOTAL TO QUOTATION				

Table 6.7.

No.	Description	Unit	Qty	Rate	£
	Dayworks — provisional				
1.	General labourer	Hour	20		
2.	Driver/machine operator	Hour	20		
3.	Concrete finisher	Hour	20		
4.	Steel fixer	Hour	20		
5.	Materials at cost $+12\frac{1}{2}\%$	Sum	—	—	1000
6.	Percentage adjustments to materials	£	1000	%	
7.	Plant in accordance with the Daywork Schedule issued by the FCEC	Sum	—	—	1000
8.	Percentage adjustment to plant	£	1000	%	
	TOTAL TO QUOTATION				

6.7 Form of Quotation

John Knapton Consulting Engineers
City Road
Newcastle upon Tyne

Dear Sir,

Quotation for: Replacement and Repair to concrete slabs

Having examined the Drawings, Conditions of Contract and Appendix, Schedule of Works and Specification for the above-mentioned Work. We offer to carry out and maintain the whole of the said Works in conformity with the said documents.

For the sum of   Preliminaries
Main works
Dayworks
Sub-total
Add 5% contingencies

TENDER TOTAL £ _____

For such sum as shall be ascertained in accordance with the conditions of Contract and Bill of Quantities returned herewith.

Yours faithfully,

Signature ....

Firm ....

Address: ....

....

Tel. No: ....

Date:..

References

Chapter I   Materials

1.1   British Standards Institution (1981). *Methods for Specifying Concrete.* BSI, London, BS 5328.

1.2   Deacon, R.C. (1988), *Concrete Industrial Ground Floors.* The Concrete Society/ British Industrial Truck Association/Storage Equipment Manufacturers' Association, Concrete Society Technical Report 34.

1.3   British Standards Institution (1992). *Aggregates from Natural Sources for Concrete.* BSI, London, BS 882.

1.4   Wright, P.J.F. (1965). *The Flexural Strength of Plain Concrete.* Road Research Laboratory, TRL, Crowthorne, Berkshire, UK, Technical Report TR67.

1.5   British Standards Institution (1982), *Concrete Admixtures.* BSI, London, BS 5075: Part 1.

1.6   British Standards Institution (1985), *Superplasticizing Admixtures.* BSI, London, BS 5075: Part 3.

1.7   Chandler, J.W.E. and Neal, F.R. (1988). *The Design of Ground Supported Concrete Industrial Floor Slabs.* BCA Interim Technical Note 11. British Cement Association, Crowthorne, Berkshire, UK.

1.8   British Standards Institution (1985). *The Structural Use of Concrete: Part 1: Code of Practice for Design and Construction.* BSI, London, BS 8110: Part 1.

1.9   Bekaert (1990). *Industrial Floors with Dramix Steel Wire Fibre Reinforced Concrete,* NV Bekaert S.A. Zwevegem, Belgium.

1.10a   Department of Transport. *Specification for Highway Works* Vol. 1, Road Pavements — Unbound Materials: Series 800. HMSO, London.

1.10b   Department of Transport. *Specification for Highway Works: Road Pavements — Concrete and Cement Bound Materials:* Series 1000. HMSO, London.

1.11   British Standards Institution (1985). *Welded Steel Wire Fabric for the Reinforcement of Concrete.* BSI, London, BS 4483.

1.12   Tatnall, P.C. and Kuitenbrouer, L. (1992). *Steel Fiber Reinforced Concrete in Industrial Floors.* NV Bekaert S.A. Zwevegem, Belgium.

1.13   (1984). *Methods of Tests for Flexural Toughness of Steel Fiber Reinforced Concrete.* Japanese Society of Civil Engineers. JSCE – SF4.

1.14   (1987). Research Committee CUR (The Netherlands Centre for Civil Engineering Research and Codes). Directive No 10, *Design, Calculation and Placing of Industrail Floors with Steel Fibre Reinforced Concrete.*

1.15   TNO (1988). TNO Institute for Building Materials and Structures, Delft, Reports B88/607 and B88/751.

1.16   Fibermesh Company (1985–1988). *Fibermesh Engineering Data Reports* Nos 1 to

7. Fibermesh Co., 4019 Industry Drive, Chattanooga, TN, USA.

1.17   British Standards Institution (1990). *Code of Practice for concrete work. Section 2.2. Sitework with in-situ and precast concrete.* BSI, London, BS 8000: Part 2.

1.18   British Standards Institution (1978). *Cold Worked Steel Bars for the Reinforcement of Concrete.* BSI, London, BS 4461.

Chapter 2   Construction

2.1   Deacon, R.C. (1988), *Concrete Industrial Ground Floors.* The Concrete Society/British Industrial Truck Association/Storage Equipment Manufacturers' Association, Concrete Society Technical Report 34.

2.2   Department of the Environment Road Research Laboratory (1984). *A Guide to the Structural Design of Pavements for New Roads.* HMSO, London.

2.3   PCF Flow Line Slab Construction, John Kelly (Lasers) Ltd. Fibrecon UK Ltd. (1992). Amorex, State-of-the-art Hardstanding Construction, *World of Concrete Europe 92, 6–9 October 1992*, QEII Centre, London. British Precast Concrete Federation.

2.4   Knapton, J. (1993). *Design of Industrial Hardstandings.* Lecture course notes, Department of Civil Engineering, University of Newcastle upon Tyne. Unpublished.

2.5   Snowdon Flooring (1994). *Technical Brochure.* Snowdon Flooring, North East Office, Houlst Estate, Walker Road, Newcastle upon Tyne, UK.

Chapter 4   Design

4.1   Westergaard, H.M. (1947). New formulae for stress in concrete pavements of air-fields. *Proceedings of the American Society of Civil Engineers* **73**, No 5, 687–701.

Chapter 5   Case Studies and Data

5.1   Meyers, B.L. and Thomas E.W. (1983). Elasticity, shrinkage, creep and thermal movement of concrete. In *Handbook of Structural Concrete* (eds. F.K. Kong, R.H. Evans, E. Cohen and F. Roll) Pitman, London, pp. 11-1-11-33.

5.2   Deacon, R.C. (1988), *Concrete Industrial Ground Floors.* The Concrete Society/British Industrial Truck Association/Storage Equipment Manufacturers' Association, Concrete Society Technical Report 34.

Bibliograpy

American Concrete Institute (1982). *Concrete Craftsman Series. Slabs on Grade.* ACI, Detroit. p. 76.

Barnbrook, G. (1975). *Concrete Ground Floor Construction for the Man on Site.* Part 1: For the site supervisor and manager. Cement and Concrete Association, Slough, UK. Publication 48.035.

Barnbrook, G. (1976). *Concrete Ground Floor Construction for the Man on Site.* Part 2: For the floor layer. Cement and Concrete Association, Slough, UK. Publication 48.036.

Barnbrook, G. (1987). *Ground-Supported Concrete Floor Slabs.* Cement and Concrete Association, Slough, UK, Reprint May 1987.

British Standards Institution (1985). *Structural use of concrete. Part 1: Code of practice for design and construction.* BSI, London, BS 8110: Part 1.

British Standards Institution (1985). *Specification for steel fabric for the reinforcement of concrete.* BSI, London, BS 4483, p. 8.

Chandler J.W.E. (1982). *Design of Floors on Ground.* Cement and Concrete Association, Slough, UK, TRA/550, p. 22.

Chandler J.W.E. and Neal F.R. (1988). *The Design of Ground-Supported Concrete Industrial Floor Slabs.* British Cement Association, Slough, UK, BCA ITN11, p. 16.

Chou, Y.T. (1983). Subgrade contact pressure under rigid pavements. *Proceedings of the American Society of Civil Engineering Journal of Transportation Engineering* **109**, No. 3, pp. 363–379.

Deacon R.C. (1986). *Concrete Ground Floors: Their Design, Construction and Finish,* Cement and Concrete Association, Slough, UK, Publication 48.034, p. 23.

Deacon R.C. (1990). Fibres for floors. *Concrete* **24**, No. 4.

Face, A. (1984). Specification and control of concrete floor flatness. *Concrete International, Design and Construction* **6**, No. 2, pp. 56–63.

Hetenyi, M. (1946). *Beams on Elastic Foundations.* The University of Michigan Press, p. 255.

Hodgkinson J.R. (1982). *Steel Reinforcement for Concrete Road Pavements.* Cement and Concrete Association of Australia, Sydney, Publication TN49, p. 16.

Hodgkinson J.R. (1982). *Thickness Design for Concrete Road Pavements.* Cement and Concrete Association of Australia, Sydney, Publication TN46, p. 16.

Knapton J. (1994). *The Structural Design of Heavy Duty Pavements for Ports and Other Industries.* British Precast Concrete Federarion/British Ports Association. Leicester/London. p. 110.

New Zealand Portland Cement Association (1980). *Design of Concrete Ground Floors for Commercial and Industrial Use.* NZPCA, Wellington, NZPCA IB0 26. p. 24.

Packard R.G. (1976). *Slab Thickness Design for Industrial Concrete Floors on Grade.* Portland Cement Association, Skokie, PCA IS 195.01D, p. 16.

Panak, J.J., McCullough, B. and Treybig, J. (1973). *Design Procedure for Industrial Slabs*

Reinforced with Welded Wire Fabric. Wire Reinforcement Institute, Austin, USA, p. 45.

Panak, J.J. and Rauhut, J.B. (1975). Behaviour and design of industrial slabs on grade. *Journal of the American Concrete Institute* **72**, No. 5, 219–224.

Pateman, J.D. (1984). Specifying and constructing flat hardstandings. *Concrete* **18**, No. 3, 7–9.

Ringo, B.C. (1978). Design, construction and performance of slabs-on-grade for industry. *Journal of the American Concrete Institute* **75**, No. 11.

Ringo, B.C. and Steenkers, R. (1982). *Industrial Floor Slabs: A Thickness Solution. ACI Computer Program Series / COM-1 (83). Design of Structural Concrete.* American Concrete Institute, Detroit, pp. 1–11.

Spears, R.E. (1983). *Concrete Floors on Ground.* Portland Cement Association, Skokie, Publication EB075.02D, p. 28.

Teller, L.W. and Sutherland, E.C. (1943). The structural design of concrete pavements. *Public Roads.* **23**, No. 8, 167–212.

Timms, A.G. (1964). Evaluating subgrade friction reducing mediums for rigid pavements. *Highways Research Board*, Washigton D.C., No. 60, p. 48–59.

Westergaard, H.M. (1925). Computation of stresses in concrete roads. *Proceedings of the 5th Annual Meeting of the Highways Research Board*, vol. 5, Part 1, p. 90–112. Highways Research Board, Washington D.C.

Westergaard, H.M. (1926). Stresses in concrete pavements computed by theoretical analysis. *Public Roads* **7**, No. 2, pp. 48–56.

Westergaard, H.M. (1933). Analytical tools for judging results of structural tests of concrete pavements. *Public Roads* **14**, No. 10, pp. 185–188.

Westergaard, H.M. (1943). Stress concentrations in plates loaded over small areas. *Transactions of the American Society of Civil Engineers* Paper No. 2197.

Westergaard, H.M. (1947). New formulae for stress in concrete pavements of airfields. *Proceedings of the American Society of Civil Engineers* **73**, No. 5. 687–701.

Ytterberg, R.F. (1987). Shrinkage and curling of slabs on grade. Part 1: drying and shrinkage. *Concrete International Design and Construction* **9**, No. 4, 21–31.

Ytterberg, R.F. (1987). Shrinkage and curling of slabs on grade. Part 2: warping and curling. *Concrete International Design and Construction* **9**, No. 5, 54–61.

Ytterberg, R.F. (1987). Shrinkage and curling of slabs on grade. Part 3: additional suggestions. *Concrete International Design and Construction* **9**, No. 6, 72–81.

Index